VIEWS BEHIND THE VEIL

Roots of Perceived Reality:
Excursions into the worldview of
physics

V. V. Raman

ARIP

i

ROOTS OF PERCEIVED REALITY

V. V. RAMAN

TABBLE OF CONTENTS

ROOTS OF PERCEIVED REALITY

FOREWORD

Ours is a technological civilization: complex, structured, machine-based, and gadget-dependent. It is wrought with a thousand wonders and dangers. It has added considerably to our heath and to creature comforts. It has also set all life on earth on a perilous course.

We also live in an age of science. But do we really? The material impacts of scientific breakthroughs have trickled down to the life of the average citizen. But the visions and insights of science, the excitement and thrills of science have not sunk into the psyche of all people in the modern world.

There is considerable misunderstanding as to what science is all about, and there is also, not unrelatedly, unbecoming anti-science grumbling, apathy for systematic science study, even an embarrassing fear of for scientific formulas. Small wonder astrology, chiromancy, numerology and "magnetized waters therapy" are thriving. Were it not for laws, witch-hunting would be practiced in some quarters even now.

But there is also a, admiration for science, caused by the feeling that the achievements of science are planes, , nuclear bombs, antibiotics TV, computers and such.

Even after four centuries of modern science, the goals

5

and framework of the scientific enterprise are but poorly understood by the average educated citizen who has little appreciation for the grandeur implicit in a scientific vision of the world, and less understanding r of what science has brought within reach of humanity.

There are some excellent popular volumes on current developments in cosmology and quantum physics. While they have been informing the general public about recent developments in physics and related conceptual breakthroughs, some of them have also created the impression that science has recognized its limits and conceded that other modes of inquiry are equally valid. Some even believe that modern science is re-discovering ancient world-views.

Many people of reasonable intelligence think that physics has at last thrown in the towel and yielded to earlier modes of speculations as to the ultimate nature of the physical world. Such conclusions result from ignoring the historical fact that philosophical quandaries and poetic extrapolations have been a feature of science in every epoch.

While homage is rightly paid to the subtle ideas and dramatic discoveries of twentieth century physics, the accomplishments of *classical* physics, i.e. the physics of the previous three centuries - which were no less subtle and dramatic - are often given only secondary importance or ignored in popular expositions of science.

The achievements of the physics of the period from the 1540s to the 1900s were considerable. Classical physics began with the recognition that our coordinates in the universe are nowhere near any central reference system, if one there be. It made us aware of the existence of planets and stars invisible to the naked eye, and even estimated their unimaginable distances. It revealed the simplicity and complexity of matter, the existence of elements never known before, radiations beyond normal human perception, and a million other things of which our distant ancestors were utterly unaware.

All this could be accomplished only by the adoption of a methodology that has observational evidence and experimental re-checking as cardinal principles. This methodology is strengthened by instruments, enriched by concepts, made precise by measurements, and penetrating mathematics. Because of all this, drastic changes were wrought in how we view the world.

Scientific awakening demolished many misconceptions about the phenomenal world; more importantly, it gave the lie to a thousand stifling superstitions, and awakened our consciousness to realms of reality never imagined by the keenest minds.

Then there are the aesthetic aspects in the explanatory framework of physics, both classical and modern, reflected in the mathematical elegance, conceptual simplicity, and logical coherence of physical theories, which are at least as satisfying as any noble work of art or elegant poetic expression.

Science, like beautiful art and melodious music, like fulfilling religion and lofty literature, is one of the finest expressions of the human spirit. It is an enriching experience to know how the world of everyday experience came to be. The labors and reflections of thousands of scientists during the past few centuries have unveiled the roots of this perceived reality. And it is of these that we will be reading in the pages to follow.

1

ON THE NATURE OF REALITY

Reality, however, has a sliding floor.

-   Ralph Waldo Emerson.

Dream and Wakefulness

Dreams are true while they last, and do we not live in dreams?   **- ALFRED LORD TENNYSON**

When I wake up in the morning I come back to my conscious state and recognize familiar things. I feel I have returned to a world of reality from which I had receded for a while during the sleeping state. My experiences during slumber, such as I recall, were fantastic and illogical and impossible, but they strike me as such only now when I compare them to the happenings in this solid and normal world to which I am accustomed in my waking hours. My dreams have included such incredible scenes as the flight of winged elephants, the walk on the ground of giant-sized fish, a performance by me at the Metropolitan Opera and a debate between Shankara and Niles Bohr on the nature of ultimate reality at which I was the moderator. It all seemed very natural and believable while I was dreaming. I did not have the wits to doubt their reality then, for, as the poet Dryden said in *The Cock and the Fox*, "dreams are but interludes which fancy makes when Monarch-Reason sleeps...."

During my waking hours I have I read the writings of philosophers on appearance and reality. Most of them are keen thinkers, but they are holding very contrary views on the subject. Each one seems to be quite sure in his or her own way about the thesis expounded. They explore questions like: What is real? What is illusion? What is imaginary? What is trustworthy? What is deceptive? What is truth? What is falsehood? These are fascinating inquiries, but only when I am reading or writing, or presenting a paper in a meeting, or arguing with opponents. Aside from making me feel good while I am playing the game, these questions don't seem to carry me too far. Yet, remote as it is from our everyday concerns, the quest for such truths is not a trivial matter. Many people take them quite seriously, and not simply as lively topics for debate and discussion. Philosophers do this in academic contexts, religious aspirants engage in it for spiritual fulfillment. Sometimes answers to such questions impinge on life and existence. In such contexts, artists express their visions of truth through painting, sculpture, music, movies, literature, and the like. Poets phrase them pithily. Some people have even given up life's normal joys and comforts in order to know what it is all about. Prince Siddhartha renounced wife and child in his restless quest for Truth and ended up initiating a whole new religion (Buddhism) that has touched and transformed millions of people over the ages. Disinterested quest for the nature of reality is no idle pursuit: it has had significant impacts on human history. Some scientists believe that they are the only ones who are intelligently dedicated to its pursuit.

I don't know the ultimate nature of Reality, but I don't feel any the smaller because of this ignorance since I have a sneaking suspicion that no human being really does. I am inclined to think that though many imagine they have unlocked the puzzle of the reason for human existence, no one knows what it is all about, much less about any reason as to why the universe exists at all. This does not mean we can never know how the world began or how consciousness arose. In fact, there have been many explanations over the ages, scientific and trans-scientific, for the origin of the world, life and mind.

I am reasonably certain that I undergo a complex of experiences every day, forming an overall picture of the world. Whether these are *real* or not, I am persuaded that I am sharing these with countless fellow humans. This commonly experienced world may be the only reality there is, although all this may be just a grand trick played on the human mind by some unknown entity, or simply by the molecules and forces in the physical world. Whatever its cause, this apparent reality is very relevant in the context of human life spans, if only because it is there, and it is unavoidable. It is the only kind of reality in which we function normally. Even if it is not all that significant in cosmic terms, the sensorially experienced world is extremely interesting and quite exciting at times.

We are all involved in this world of conscious experience in our own ways, and we function in it with some notions, clear or unclear, articulated or not, as to the nature of truth and of the physical world.

To normal human beings, the world has two aspects: the collective and shared phenomenal world, which is external, and is fascinating as a subject matter for study; and the deeply personal experiential one which involves feelings and emotions associated with the course of one's own life. These two aspects are not altogether unconnected, since both arise from the sensory perceptions which activate us as conscious entities.

V. V. RAMAN

Perceived reality

Habet cerebrum sensus arcem: The brain is the citadel of the senses. - THE ELDER PLINY

Most people are endowed with functioning faculties of perception during a significant period of their lives. We become aware of the world through eyes and ears, nose, tongue and touch. These channels of perception transform physical stimuli into sensations, creating pleasure, pain, and impressions. The glory of the sunset and the scent of perfumes, magnificent music, delightful tastes of food, and the softness of a cat's fur: we have all experienced such things, thanks to neurons that keep firing in our brains as and when occasions arise.

There is neither music nor color, neither odor nor sweetness in the crass world of matter and energy. Out there in the world beyond our bodies, there are only matter chunks and vibrations, silent and senseless, dark and dreary. The transformation of all this into an astounding world of beauty and fragrance and enjoyment is brought about by the brain, perhaps the most wondrous entity ever to have evolved in the cold cavity of the universe.

The stimuli also generate some abstract entities (thoughts) in our brains. Each one of us is, in Pascal's phrase, is *un roseau pensant:* a thinking reed. The brain

13

also provokes feelings and emotions: joy, sorrow, exhilaration. The perceptual inputs convey information about the world. The totality of all experiences, impressions, and information obtained through the normal channels of sensory faculties constitute the world of *perceived reality* (PR).

"Science," Plato said long ago, "is nothing but perception." It is, more exactly, the careful, rational, and consensus-based interpretation of commonly shared perceptions. For ultimately, nothing is science if it not related to some aspect of perceived reality, if it is contrary to reason, and if there isn't a consensus among scientific investigators as to its actual or potential validity. These are necessary, if not enough conditions for what constitutes science.

Categories of brain-modes and the challenge

And here we wander in illusions. Some blessed power deliver us hence! - SHAKESPEARE (*The Comedy of Errors*)

Another characteristic of the human brain is its capacity for logical reasoning: another intangible wonders in the universe. As far as we know, this process naturally occurs only in the human brain, though human ingenuity has invented devices, called computers, where processes very close to logical reasoning occur at stupendously faster rates. It would be rash to describe human beings as rational animals or logical creatures. More exactly, we are capable of logical thought. Logical thinking is akin to walking on a perfect straight line: something that, in principle, we all can do, but which, in practice, we do not always do.

The human brain is also capable of creating a variety of misleading impression-modes. Because of an interplay of events and circumstances, distorted aspects of perceived reality are sometimes registered in the brain. These are called *illusions*. The worldviews arising from illusions may be called *delusions*. Sometimes, even in the absence of external stimuli the brain may generate its own world which is utterly unreal. We call these *hallucinations*.

The brain can also go beyond perceived realities and generate magnificent worlds which widen and enhance our experiences. These worlds are seen only through the mind's eye. They may, and often do, stray away from logical constraints, and they can also have great charm and meaning and power. This is the realm of *imagination*, which is also at the root of the creative potential of the brain, a capacity that has added immensely to the richness of the human experience, at least as much as our ability to reason and analyze the tangible world.

The brain is also capable of constructing its own tools for describing perceived reality, which are indispensable in our efforts to interpret the world. It is thus no surprise that the task of understanding the world is a very complex one. We need to sort out illusions and hallucinations from the normal modes of perception, utilize our capacity for imagination without being misled by it, erect useful and appropriate mental constructs to describe the world effectively. In the process of doing all this, we also need to be sensitive to our deepest emotional needs and cultural heritage, without being overwhelmed or constrained by them.

Human experience as máyá
máyám tu prkrtim viddhi: Know, however, that the world is illusion.

- SHVETAASHVATARA UPANISHAD

Ancient philosophers in India propounded a vision of reality that is relevant in our interpretation of the world. The essence of ultimate reality, they declared, is normally veiled from our apprehension because of several constraining factors that are imposed on matter and mind. The result of all this is the creation of *máyá*, a world of illusion to which normal human experience is inevitably subject.

Since máyá is a veil that screens the true nature of reality, it is often regarded as something negative. But máyá is not necessarily evil and hurtful, once its innate illusory nature is recognized. On the contrary, máyá serves us very well in the course of our lives and is in fact at the root of many of our enjoyments, institutions, intellectual exercises, and societal interactions. Certain aspects of máyá have even helped us gain a deeper understanding (or at least visions of higher categories) of reality.

Thus, one may say that art is *aesthetic* máyá. A great painting is an illusion which transports us to a grand experience, as indeed is any work of art. As Goethe said, "The highest problem of every art is, by means of appearances, to produce the illusion of a loftier reality."

Literature may is *emotional* máyá. It paints imaginary people and unreal events, kindles our longing for ideals, allows us to express our capacities for anger and compassion, and helps us in our search for the deeper truths relating to human nature and to the human condition. We know that no page in a novel is true, yet great literature touches us profoundly and stirs our emotions.

History, in theory, is a record of major events that occurred in the past. In most instances, however, it is a narration of events that did not happen the way they are reported. The goal of history often is, intentionally or otherwise, to make a people feel good about their past. History is often, if not always, *patriotic* máyá. It serves to boost the self-respect of a people. This is why history becomes a difficult subject in multicultural societies.

Mythology is another grand máyá that has played a major role in human history. It speaks of gods and demons that never existed, in terms and language that are inspiring and fascinating. Mythologies make a people good, creating in them, like history, the impression that their past was graced by majestic beings, by powerful heroes and heroines who fought for the good and defeated the evil. Myths are thus *inspirational* máyá.

Religion may be seen as *spiritual* máyá, creating the impression that we truly know about the nature and attributes of a divine principle. But it gives meaning and purpose to life, and more importantly, it has been, and continues to be, the source of solace and psychological comfort to countless millions.

In mystical experience, one sees visions of a reality that is probably the result of abnormal brain biochemistry. Mystical experience is a powerful *mental* máyá which has nevertheless instigated affirmations that have transformed the lives and world views of many individuals and groups.

Philosophy is *speculative* máyá, a play with concepts and ideas, creating systems to interpret in broad terms the nature of human thought and worldviews. With careful analysis and argumentation, however, every system of philosophy crumbles down.

Finally, we have science. Its goal is to reveal the nature of physical reality. But science too is a kind of máyá. The history of science shows that many past theories and explanations of science were wrong. Yet we have been able to do many things with inventions based on scientific knowledge. Science is thus *fruitful* máyá. Its framework provides us with intellectual satisfaction, and with its results we can construct a great many things to satisfy our need and greed. Likewise, recognizing the chasm separating theory from practice, one may say that democracy and communism are political máyás.

Thus, not just individual lives but human civilization and institutions too are based upon máyá of all kinds. We are condemned or blessed to lead a life that is both distorted and enriched by the variety of máyá.

In all this we note that many things that enrich us and give us enjoyment are not exactly what they appear to be. To probe into the deepest levels of truth does not mean we should reject or despise the benefits of its superficial aspects.

Scope and limitations of perception

The senses collect the surface facts of matter...When mind acts, it is knowledge.

- RALPH WALDO EMERSON

The universe is a museum with a splendid collection of things and processes. As noted, our faculties of perception are the most remarkable systems we have recognized thus far in it. There is nothing in all the world that we have been able to uncover, here on earth or elsewhere, which is more spectacular in scope and capacity than the little muscular aggregation that we all possess under our skulls.

What a marvel, this concentration of chemicals we call the brain! It transforms silent photons into chromatic splendors, changing pressure waves into glorious music, rough edged molecules into pleasing odors. It brings to our awareness the existence of distant entities and ancient happenings and future possibilities. And it can also reason mathematically.

But this truly wonderful instrument has its limitations also. For one thing, it cannot discern very feeble stimuli. There are threshold levels under which no light will be detected, no sound will be heard, no smell will be known. Nor do our sensory faculties pick up every kind of signal the world. A great many waves go reckoned. A great many things go unrecognized because our sensory faculties just do not respond to them in any perceptible way. As a result, reality we perceive is only a partial mapping of all there is.

Science tries to compensate for these limitations by devising instruments. Some instruments magnify the feeble signals to make them detectable. Thus the telescope enables us to see faint sources of light which are virtually invisible to the naked eye. On the other hand, radio-telescopes put into evidence radiations from space that, no matter how intense they are, can never be detected by the naked eye. Scientific instruments enhance the level and expand the range of the PR to which our unaided faculties are normally attuned.

It is important to be clear about this: Our awareness of the physical world has been enhanced immeasurably by instruments that scientific investigators have devised during the past few centuries. This is consciousness-raising in the context of reality; and it is very different from the psychedelic hallucination that chemical addicts rave about.

2

THE GOALS AND FRAMEWORKK OF SCIENCE

Men love to wonder, and that is the need of our science. **– R. W. EMERSON**

----

*Was man nicht versteht, besitzt man nich; What one does no*t understant, one does not possess. **– GOETHE**

Now we come to what physics is all about: Physics is a concerted attempt by human minds to seek the underlying order in terms of which we can better understand and appreciate the nature of perceived reality such as it is.

The sun does not really rise, not the rainbow a bow that arches from sky to ground; the oar does not really bend when partially immersed in water, and neither are stars fixed in high heaven. We know of a thousand appearances that have been exposed to be what they really are. We may call it the deception of Nature or Nature's reluctance to unravel herself to the superficial onlooker. The perceived world mimics us humans, for do not people all too often put on appearances that hide

22

the thoughts they harbor or the feelings they experience? However, with experiments, insight and clever questioning, one can fathom the deep-down secrets beneath a feigning face of Nature.

If the world may be compared to a Being, what we see is a veiled being, fully clothed to impress and entertain, but her muscles and structure, its features and subtleties remain teasingly hidden from our normal view. We need to do a lot of wooing and cajoling and tinkering to persuade Nature of incredible power and beauty to undress and reveal all her magnificence and innate secrets. When she is thus stripped and we behold its pristine glory, there is an astounding simplicity about her, a beauty and balance that words of human language can barely express. We need to invoke the magical modes of mathematics for this.

Simply, put, this is what physics, and more generally all science, tries to accomplish: to unravel Nature to know her steadfast behavior, and to fathom her deepest thoughts, so to speak. For there is one thing we have come to know: Nature is unwavering in her functioning, unerring in whatever she does.

Though one may look upon the order and patterns in Nature as arising from blind and brute forces, it is equally persuasive to consider all the measurable and mathematical aspects of perceived reality as resulting from some supreme intelligence directing the course of events based on cosmic calculations. Even if this be not the case - and there are die-hard scientists who would vehemently object to such metaphors - little is lost and much aesthetic satisfaction id gained from such a poetic vision, if that is all it is. As the mathematician Joseph Lagrange said, it is a pretty hypothesis which explains so many things (*jolie hypothèse qui explique tant de choses*).

The human mind for the physicist

Vivida vis animi pervicit, et extra longe flammantia mœnia Mundi: The lively force of the mind has broken down all barriers, and has made its way far beyond the glittering walls of the Universe. - LUCRETIUS

Because it is the human mind that does the searching, we are perforce obliged to function within the framework and possibilities of the mind.

Here again, a fundamental question arises: What exactly is the ultimate nature of the mind? We have centuries of loaded answers to this question; if and when one has the time or interest, one may look into what countless thinkers, from Upanishadic seers and Plato through René Descartes and David Hume and Immanuel Kant and Sigmund Freud down to current metaphysicians and psychologists have to say on the subject. Like pleasure and pain, we may not know what exactly mind is, but we do know it is a perennial aspect of normal conscious life.

The capacities of the human mind are limitless: thought and recognition, analysis and questioning, concept-building and mathematizing, and much more. How do these capacities arise? Are they simply the inevitable consequences of the physical structures and the chemical properties of the human brain, as many present day scientists aver, or do they arise from some external mysterious source imposed upon us, like the image of the moon in a pool of clear water? We do not know for sure even though arguments, persuasive as well as fractured, have been presented for both points of view by able thinkers over the centuries.

Physics, in its more serious and laboring moments, is indifferent to the debate. Philosophers theorize on the nature of the mind, theologians preach about it, psychologists analyze it, and neurophysiologists explore it in terms of cerebral axon-potentials. Physics accepts the mind such as it is and exploits its rational capacities to provide self-consistent principles in terms of which the perceived reality may be framed. It seeks to uncover universally acceptable patterns which would throw light on why Nature behaves the way she does.

How reliable can be results derived from a system based on indifference to the innate nature of its most fundamental tool? This is a legitimate question, and it has been considered by many. There is a story to the effect that when the Greek philosopher Zeno proved by reasoning that there can be no motion, Diogenes simply stood up and walked without giving a counter-argument. He did not give any proof that there was something wrong in Zeno's proof. Likewise, while others vigorously argue about the limitations of the mind and the inherent difficulties in deciphering the ultimate nature of reality, working scientists go ahead and bring the world within reach of the discerning mind. Whether scientific truths are permanent or passing paradigms, physicists can't be bothered. What matters is that their efforts do produce tangible consequences.

Planes of interaction

A man, always studying one subject, will view the general affairs of the world through the colored prism of his own atmosphere. **- BENJAMIN DISRAELI**

Recall that much of our apprehension of the world is máyá. We interact with perceived reality on different, sometimes overlapping, planes.

We are *physical* beings. When we consume food, we are interacting with the world at the physical level. Whether poet, philosopher or physicist, saint, singer or sportsman, one needs to eat and drink to be able to engage in any activity. The body needs energy for functioning, it must be kept in working condition, and it also craves for titillation. We not to savor food and drink. We walk and run, play and relax, and indulge in physical pleasures. Interactions on the physical plane serve these purposes.

Then there is the *emotional* level of interaction, especially in our dealings with fellow humans. On this plane arise feelings of love and hatred, anger appreciation, kindness, compassion, and so forth. Our existence as human beings is enormously enriched by interactions on the emotional plane. They become relevant in relationships, and they enable us to function in the world with sense and sanity. Bereft of feelings and emotions we cease to be humans. Feelings also enable us to apprehend an aspect of perceived reality in a special way. There is truth in the old saying, "Seeing is believing, but feeling's the naked truth," for what we feel in our very core is what we take to be the truth.

We interact with perceived reality at the *aesthetic* level. This involves the elevated experiences we drive from visual beauty, auditory delight, and the magic of words. The splendor of the serene sunset, the majesty of mountains, the paintings of the masters and the sculpture of creators who transform mute marble into static life, enchanting music and birdsong, rhythmic poetry and vibrant rap, movies and plays, all these are among the countless ways in which we interact with perceived reality on the aesthetic plane.

We interact with the world on the *moral* plane where values come into play. Here arise questions of justice and fairness, of right and wrong, of good and bad, and such. Explicitly or otherwise, we function in a framework of values. Our linkage to this plane often determines our actions and attitudes, our interests and behavior. As members of a society we cannot afford to be indifferent to this plane.

Another plane of interaction is the *spiritual*. Here we experience awe at the majesty of the universe, and marvel at our uniqueness as conscious beings. We focus our mind on an unknown grandeur, on an abstraction that is at once magnificent and mysterious, all pervading and everlasting, and we yearn for a sublime connection with the cosmic whole. This is prayer, this is meditation. We try to give forms and names and attributes to an unknown mystery.

Finally, there is the *intellectual* plane. Thinking and reasoning dominate this mode. Logic becomes dominant. We seek consistency and reference systems. We measure and manipulate. We observe and record. We carefully comb our findings to spot misleading intrusions. We try to prove and search for mistakes in proofs. Anything that deviates from reason is discarded, anything that affronts logic is laughed out of court. It is on this level that the world of science operates. If our goal is to explain the world in intellectual terms, then this is the plane on which we ought to work.

It would be absurd to claim that interaction on one plane is more important than another. All these are aspects of máyá, and no one mode is more valid than another in an absolute sense. As human beings we can, we do, and we must, interact with perceived reality at all levels. These planes of interaction are not impermeably separated, but we may be thrust into awkward corners if, while functioning at one level, we invoke the tools or methodology of another.

Role of mathematics
The book [of Nature] is written in the mathematical language... without whose help it is impossible to comprehend a single word of it; without which one

wanders in vain through a dark labyrinth.
- GALILEO GALILEI

An interesting feature of perceived reality is that it is amenable to quantitative descriptions. We can associate numbers and mathematical relationships among the elements and events we recognize in the world. It is difficult to say if these are intrinsic to the world or impositions by the human mind on what is observed. Be that as it may, we can describe the world through formulas and equations.

The mathematical mode is a key to a deeper understanding of the roots of perceived reality. Centuries of earnest and intelligent efforts to unravel the nature of the physical world produced only a modest harvest compared to what has been accomplished in the last few centuries, largely because of just two or three factors: and one of them is the transcription of the experienced world into the framework of mathematics.

Translation into mathematics is not the same as translating into any of the usual languages. For mathematics is a language only in a metaphorical sense, like music and painting. Like art and music, mathematics transports us to a different realm of experience, and it also endows us with a penetrating power in our grasp of perceived reality.

Like the telescope that has brought within our visual range aspects of the universe of which we cannot otherwise have an inkling, mathematics has revealed to us the structure and subtleties in the roots of perceived reality that speculation or meditation can ever unravel. Mathematics is more than a tool, more than a nutcracker which enables us to break open a nut, or a knife with which we can slice an apple. Take away these tools, and we can still crack the nut with brute muscular force and bite the apple with good teeth. But take away mathematics, and we can never move into the marvels of the microcosm. Mathematics is more like the space-ship of the astronaut who cannot plunge into the depths of space without it. The physicist can't delve into the ultimate substratum of matter and energy without mathematics. It is an indispensable faculty of the human mind in unearthing the roots of perceived reality.

This is what makes physics an esoteric discipline. This is what frightens the non-initiate, for there are too many things that are abstruse and abstract in the compact formulations of physics. The symbols are strange, the reasoning as incomprehensible as epics in an unfamiliar tongue, and the formulas are too tightly packed for immediate comprehension. It takes time and training to understand the mathematical jargon in which physicists describe the world.

Without boarding a rocket, we can get excited by descriptions of the lunar landscape given by Neil Armstrong. In like manner, there is a good deal about the physicist's findings that the interested reader can enjoy without plowing through computations, calculus, and coordinate geometry. This is the conviction on which these pages are composed.

Role of observation

Armando: How hast thou purchased this experience?

Moth: By my penny of observation.

- SHAKESPEARE (*Love's Labor's Lost*)

There are not many people who haven't *seen* the moon. But there are not too many who have *observed* the moon. Observation is much more than seeing with interest and enthusiasm. Observation involves sustained interaction with an object or an event in which the mind plays an active part. When one observes something one takes careful note of every aspect of whatever is within reach.

Philosophers, poets, theologians and others make many general statements about the world and its origins, about life and beyond, about truth and the nature of reality. Many of these are insightful, interesting, and thought-provoking. In the methodology of science, however, before one makes a general statement about perceived reality, one tries to make general statements about its specific aspects.

For this one needs to study the various aspects of the world in all their details with great attention. Scientific observation implies the amassing of as much information as possible on whatever is studied. The data are largely in the form of notes and numbers, some of which may be transformed into tables and graphs. All these are then organized, classified, analyzed, and interpreted. The conclusions one draws from the data of observation are often in the form of generalizations about the specific matter that is studied. Such generalizations are called (empirical) laws.

Observation is the first step in bringing any aspect of perceived reality within the framework of science. It does not always require instruments. If a statement about any aspect of the world is not connected to the data of observation, it is of little relevance to science, however pleasing or interesting it may be.

Observer and objectivity
Fortunately for serious minds, a bias recognized is a

bias sterilized.   **– EUSTACE Haydon**

One goal of science is to obtain *objective* appraisal of the world. This means an understanding that is independent of the persons who do the investigation, and of the circumstances under which knowledge is obtained. Striving for objectivity implies a serious commitment to eliminate all personal factors and prejudices in our efforts to unravel the nature of any aspect of perceived reality. As stated in his *Physics and Philosophy: The Revolution in Modern Science* (New York: 1962, p. 82), "Every scientist who does research work feels that he is looking for something that is objectively true." Objectivity also implies independence from human minds.

Philosophers have analyzed the notion of objectivity in science extensively. [e.g. Karin D. Knorr-Cetina, *The Manufacture of Knowledge: An Essay on the Constructivist and Contextual Nature of Science*, Oxford: 1981.) In the modern scientific tradition, René Descartes was one of the first to articulate this criterion for scientific knowledge when he introduced the notion of a world being made up of a *res cogens* and a *res extensa* by which he meant a subject which did the observing and an object which was being observed.

This approach has been sometimes condemned for

initiating a dichotomy which divides the integral whole into exploiting Man and exploited Nature. This led to the industrial revolution, environmental pollution, and ultimately (the impending) extinction of our species.

One also objects to the Cartesian separation on less deprecating grounds. For one thing, if there can be no science without observation, then there cannot be any science without an observer either. And if observations are made in the language and conceptual framework of the observer, then in what sense and to what extent can the knowledge thus acquired be regar5ded as objective? Thus the very goal of science, namely an objective description of the world becomes logically untenable.

One answer to this is that by *objective* one means an understanding that is independent of who (i.e. which person) does the observation. In other words, scientific knowledge is supposed to be universally acceptable. It must be appealing and convincing to one and all who examine the phenomenon with the same degree of interest, attention, and demand for consistency. Scientific objectivity simply means collective (human) subjectivity.

Even granting this, there is a deeper sense in which Descartes' subject-object schism will have to give way. While it is possible in our everyday world to objectify the world to make observations without interfering with the observed, this is impossible, even in principle, in the microcosm. When we investigate processes in the world of atoms and electrons, we inevitably interfere with what is being observed to the point that the results of our observation cease to be subject-independent. This leads to the suspicion that deep down there is an inseparability, hence an interconnectedness, between the observed and the observer.

Experimentation

We will answer all things faithfully.

- SHAKESPEARE (*Merchant of Venice***)**

Since ancient times Investigators have tinkered with things in their efforts to understand them. When this is done systematically with care and instruments with the goal of eliciting detailed information from the silent world of physical reality, it is called an experiment.

The physical world is one large interconnected whole whose myriad parts function in harmony, and yet in such grandiose complexity that it is impossible to embrace them in their totality, except when one is functioning in a meditative mode. Science, by its very nature, is analytical: a piece-meal probe into perceived reality. What this means is that if we wish to dig into the roots of some aspect of perceived reality, we need to concentrate our attention on it and examine it *in extenso*.

An experiment in science may be looked upon as a personal interview that an investigator arranges with a well-defined aspect of the natural world. It is pre-arranged and intentionally set up. It entices the desired aspect of the world with precise instruments and measuring devices. It includes specific questions that the experimenter asks of whatever is being interviewed. The scientist is trained in the language and format of the questions to be asked, for one cannot conduct a scientific experiment without a well-defined conceptual framework. The experimenter must know how to interpret the answers given. The interpretation of the results of experiments often advances our understanding.

Sometimes the results of experiments are clear, sometimes hazy. Sometimes they reveal a hitherto known or unsuspected related aspect of perceived reality.

This method tends to be offensive to those who are not inclined to science. Recall what Emily Dickinson said about the botany teacher (*Poems*, Pt. ii, No, 20.):

> I pull a flower from the woods -
> A monster with a glass
> Computes the stamens in a breath,

And has her in a class.

This perfectly justified reaction arises when one tries to interact with the world in two different planes simultaneously, or when one assumes that one mode is intrinsically better than another. If we wish to derive the aesthetic joys the flower affords, we must adopt one mode; if we wish to understand intellectually certain aspects of the flower we need to interact with it differently: we must analyze it and experiment with it.

Explanations in science
I wish he would explain his explanation.

- **GEORGE GORDON BYRON**

A primary goal of science is to account for whatever is observed. The need for such an accounting arises from the implicit belief that nothing happens by itself, without a definite cause. Once the cause of a thing or event is determined whatever one is investigating seems to have been explained. One feels one has understood the aspect of perceived reality that is under study. Explanation consists in creating the impression that we understand something.

There are levels of explanations, depending on the sophistication of the person who gets the explanation. If a child is told that music comes from a radio because there is little man in there who is singing, it may be satisfied. If an educated person is told that leaf is green because it contains chlorophyll, he/she may be satisfied, even though this technical word simply means *green leaf* in Greek. On the other hand, the physicist's explanation in terms of quantum mechanics of why water is stable will be understood only by those who are familiar with the language and framework of quantum physics.

In all instances of explanation, there is a conceptual connection between two or more elements. In other words, the explanation of a thing A will be in the form: *A is there because of something called B*. The connection between B and A may be either direct or direct; it may be obvious right away or tortuous and not so obvious.

Sometimes there may be two or more explanations for the same perceived reality. An explanation offered by one generation of scientists may be superseded, improved upon, or discarded by another. The history of the scientific quest has two principal dimensions: (a) the gradual uncovering of newer aspects of perceived reality; (b) the changing series of explanations offered to various aspects of perceived reality.

Science is applauded for the first of these, for new discoveries imply new knowledge. It is criticized for the second, because new explanations mean that the science of one period has been giving wrong understandings and misleading interpretations to people of a previous generation. This, declare the critics of science, makes scientific knowledge suspect, unreliable, and by no means a claimant to truth.

Here it is important to recognize two things: First, though scientific explanations are not necessarily correct in the *absolute* sense that they will never be replaced, they are the most valid ones we have at any given period: i.e. they are *the most rational, coherent, and consistent answers one can give in the context of currently available data on a given phenomenon.* This, by definition, is the nature of scientific truths, and no other system has been able to offer an intellectually more satisfying or reliable body of truths. Those who fault science for its inability to provide hundred per cent certitude generally do not give better explanations.

Laws of nature

Laws of Nature are God's thoughts thinking -themselves out in the orbits and the tides. - C. H. **PARKHURST**

One mode of scientific explanation is by relating the various elements of perceived reality to one or more *laws of nature*. A law of nature is a concise statement of a (suspected) behavior pattern of the physical world in a specified context. In other words, whatever is observed may be understood to be consequences of one or more of the laws which govern the functioning of the world.

A law of nature may be discovered by observations alone: from the data of observations one may detect patterns of behavior of the physical world. For example, one may deduce from observations that the pressure exerted by an enclosed volume of gas increases when the gas is compressed (i.e. when the volume is decreased). This reveals the general behavior of gases, and is therefore taken as a law of gases. Similarly, by observing the positions of images formed in mirrors relative to the object positions one may state the rules by which light is reflected from a mirror. This leads to the laws of reflection. By examining extensive data relating to the orbits of planets, Kepler concluded that planets go around the sun in elliptical orbits.

The term *law* (of nature) came into use in science because of an anthropocentric analogy. The thinkers of the seventeenth century believed that the broad patterns of behavior of nature were like human behavior which is governed by (human-made) laws. The implication was that there is a higher principle (the author of the universe) which has decreed these laws. In Newton's words, "The wonderful arrangement and harmony of the cosmos originate in the plan of an almighty and omniscient being." We must note, however, that while human-made laws may be modified or broken, the so-called laws of nature cannot.

What is remarkable is that the wondrous variety in the world, all the magic of this colorful and complex world, all the obvious order and apparent randomness, arise from and in accordance with mathematically precise rules and regulations (which we call *laws*). When a law of nature is apparently broken, we call it an anomaly or a miracle, though the greater miracle is that laws operate at all in an apparently mindless universe.

Hypotheses and theories
... those things which neither can be demonstrated from the phenomenon nor follow from it by argument of induction, I hold as hypotheses. - ISAAC NEWTON

There is another way science explains aspects of perceived reality. It constructs in our mind models of not-directly perceived aspects of reality, and drawing consequences from such models. For example, if we see an apple fall from the tree, we say that the earth is pulling the apple down, though we do not see the actual pulling. Such a conceptual picture of any aspect of how (we imagine) the world behaves is called a hypothesis.

Thus, a hypothesis is a statement about the roots of some feature of perceived reality. Science is willing to consider it, provided that it leads to logically derived conclusions that correspond to significant aspects of perceived reality. A hypothesis becomes interesting only from the consequences that follow. Then its value increases or diminishes.

The logical (and mathematical) derivation of the consequences of a hypothesis (or set of hypotheses) constitutes a scientific *theory*. This word has different connotations. Thus one speaks of Plato's theory of ideas, Kant's theory of knowledge, Cantor's theory of numbers, Hobbes's political theory, and the impressionist theory of painting. In all these contexts *theory* means something quite different from its connotation in the phrases like *Bohr's theory of the hydrogen atom* or *Einstein's theory of relativity*.

Theories are the explanatory structures in terms of which science tries to understand perceived reality. If PR is compared to the branches of a tree, hypotheses are the roots, and theories are the trunks which connect the unseen roots to the experienced branches through the (logical) trunk.

Anyone may formulate hypotheses in science. But if

they are to acquire recognition from the scientific community, they must pass the test of the *so-what criterion*. According to this, when a hypothesis is proposed, it must answer the question, *so what*? If the answer to this question leads to statements that have been or can be verified or falsified by observation and experiment, the hypothesis is accepted. If not, it loses a place in scientific discourse. When a hypothesis is stated without consideration of its consequences and taken as valid even if the derived consequences are in blatant contradiction to what is observed, it becomes a *dogma*.

Curiosity and excitement

Curiosity is one of the most permanent and certain characteristics of a vigorous intellect.

- SAMUEL JOHNSON

The scientific quest is characterized by an insatiable curiosity to uncover the roots of perceived reality. There is no limit to the range of inquiry. Curiosity may kill the cat, but it also kindles the scientist. Not all thinkers have regarded curiosity as a virtue. St. Augustine was told that God had created hell especially for the inquisitive.

. Many have warned that in the long run scientific inquisitiveness would prove to be more harmful than beneficial to humankind. In the sixteenth century, Agrippa von Nettesheim, who described knowledge as the "true plague that invades all mankind," warned that "the pursuit of the sciences is so dangerous and unpredictable that it is far safer to be ignorant than to know." *On the Uncertainty of our Knowledge*, Ch. I.)

Nevertheless, curiosity about the world has been there in all epochs, and the excitement of discovery is as ancient as the human spirit. The legendary *Eureka!* cry of Archimedes, upon discovering the principle of buoyancy of fluids, is symbolic of the intellectual thrill experienced by the scientific discoverer. Recall this passage from the biography of Madame Curie:

"Powerful and tranquil, he (the professor) ventured into the most tenuous region of knowledge, he played with numbers, with the stars; and as he was not afraid of imagery; he pronounced in the most natural tones, accompanying the words with the easy gesture of a great property owner, 'I take the sun, and I throw it...' The Polish girl on her bench smiled with ecstasy. Under her great swelling forehead, her gray eyes, so pale, were illumined with happiness. How could anybody find science dry?"

Not every student may experience such joy in the classroom. Even thousands of people involved in scientific projects may not always feel this way. Yet the essence of doing science is here: a thrill in becoming aware of the roots of perceived reality. It may not be one's own discovery. But unless one experiences a personal joy, almost mystical, even in the learning process, upon viewing a root of the apparent world, one is not fully participating in science.

Skepticism in science
Non menche saver, dubbiar m'aggrata.
No less than knowledge, doubt charms me.

- DANTE ALIGHIERI

Skepticism as a philosophical attitude towards the human capacity for acquiring definitive knowledge has a long and prestigious history. Many have subscribed to the view that it is impossible for the human mind to comprehend the world fully and correctly by means of reason or whatever other mode. In simple terms, skepticism is the tendency to doubt everything.

Doubting is essential for seeing things more clearly and for acquiring new knowledge. As Housman put it,

> Human minds so move about
> Only if fenced round with doubt;
> Only if denied their grasp
> Gain the everlasting clasp.
> Only streams which fettered be
> Fret their way at last to sea.

However, the scientific mind is skeptical, not in the philosophical sense of contending that correct knowledge is impossible, but in the etymological sense of the word: *skepticism* is derived from a Greek verb which means *to consider, to examine*. This implies in the scientific context that any statement about the world should be subjected to severe analysis, criticism, and test before it is allowed public appearance, not to mention acceptance and recognition as a valid world view. Older views of the world are given similar treatment also. In other words, scientific skepticism is *à priori*, i.e. before conclusions are drawn; whereas philosophical skepticism is *à posteriori*, i.e. it is related to the conclusions themselves.

Skepticism is inculcated in the teaching of science through laboratory courses. One of the aims of laboratory instruction is to convince the student that what he/she reads in a book or is told in a lecture is not to be accepted as such but must be confirmed and verified. Implicit in lab courses in schools is to teach students to verify things on their own.

Authority in science

Do not believe in something simply because it is tradition and it is old. Do not believe in anything on the mere authority of myself or of any other thinker.

- GAUTAMA BUDDHA

This injunction was given more than 2500 years ago in the context of religious truths. It may well be adopted as the motto of the scientific enterprise. Most religions are based upon the pronouncements of individuals or utterances in holy books. These are to be accepted unquestioningly because they are regarded as revealed knowledge, i.e. visions emanating from a higher mystical source. This is a perfectly valid position to take when one is pursuing a (particular) religious path.

All through history some thinkers have been unable to accept this principle; some have rebelled against it, even at risk to their own lives. While we may commend the independence of thought of such individuals, in the religious context such an attitude refuses to play by the rules of a game. If one disagrees with the rules, perhaps one ought to switch games rather than try to modify the rules.

This is what science is all about. It is an attempt to play the game of discovering the roots of perceived reality without conceding the possibility of revealed truths to charismatic individuals. The idea is not that we should reject what others have said or taught, but that nothing is necessarily true simply because it comes from a revered authority. More importantly, reliance on received wisdom is seldom the sure path to new knowledge.

There are authorities in every field of science. These are individuals who, by virtue of serious study and years of research, have mastered the multitudinous details in the field, and have made solid contributions to it. An authority in science is respected and relied upon for depth of understanding; but the authority is never regarded as infallible. Scientific authorities are referred to as *experts*. The real authority in science is not an individual, but a body of experts in the field. This body is not elected, nor vested with special powers. Even presidents and members of scientific academies do not, in principle, have power over the common practitioners of science.

Challenge to venerated truths is not just permitted, it encouraged in science. A research grant may be regarded as a financial incentive not only to find out something new but also for overthrowing something old.

Humility in science
We really know very little, and we are all fallible when facing the immense difficulties presented by investigation of natural phenomena.   -

CLAUDE BERNARD

Perceived reality is abundant in its manifestations, infinite in scope. Even the best of human minds can hope to grasp only an infinitesimal part of it all. Therefore, another ingredient of the scientific attitude is humility: it is a humility that recognizes not only one's own limitations, but those of others as well. In other words, the scientific worker, while he/she is engaged in research, takes it for granted that the investigations of fellow workers, no matter how experienced or reputed, are quite capable of making mistakes, grievous or small. Therefore, it is incumbent upon the practicing scientist to view with very critical eyes the work of others in the field. This suspicion about the correctness of others' work is manifest as sharp questions and technical challenges. In the scientific world, this is regarded as compliment, not as criticism.

Thus, progress in science occurs, not by the efforts on just one individual, not even by a single nation or generation, but by a collective world of many who, by mutually correcting one another's inadequacies and by taking advantage of one another's results, constitute a kind of superpesonal mind. This superpesonal mind, strengthened by the collectivity and rendered relatively immune from major blunders by the vigilance of its parts, attempts to disentangle the roots of perceived reality. That is why science is a collective enterprise. That is why the probability of error in a scientifically held proposition is far less than one made by a single personality, however wise or prestigious.

But even such carefully and collectively formulated principles could be mistaken. What Neville de Chaussée said in a different context, is very applicable as regards scientific truths. *Quand tout le monde a tort, tout le monde a raison*: When everyone is wrong, everyone is right. Currently accepted scientific truths are taken to be right only because all scientists hold the same view on the matter. Another generation may prove them wrong.

The ideal and the real

And is this Yarrow, this the stream
of which my fancy cherished?
So faithfully a waking dream,
an image that hath perished! - **WILLIAM WORDSWORTH**

The heroic picture painted in these sections must be qualified. Eventually and inevitably objective evaluations determine the success or otherwise of ideas in science. But this is true only generally. There are specific instances where non-scientific factors direct the

51

course and currency of scientific theories. People who, by virtue of contributions and repute, hold prestigious positions and purses have advanced their own ideas and results. They have discouraged what may seem to have the potential to devalue their ideas.

This is not surprising, because science is a human enterprise, subject to human failures and foibles who are prone to succumb to personal benefits and glory. Robert Merton listed some of the shortcomings from which scientists are not immune. Here he included such things as, "contentiousness, self-assertive claims, secretiveness... false charges of plagiarism, even occasional theft of ideas and, in rare cases, the fabrication of data..."

4

SPACE: THE EXPANSE OF VOID

Space is a conception of many aspects, and it has arisen - under various names, appellations, and descriptions - in different areas of cognition and knowledge: in cosmology, physics, mathematics, philosophy, psychology, and theology. - **SALOMON BOCHNER**

The vast stretch beyond
But how can finite grasp Infinity? - JOHN DRYDEN

One summer night I was outdoors and in silence I stared skyward. I pictured myself in an all-powerful vehicle that sailed indefinitely on and on, and in my flight of fancy I took off far above the clouds and into the expanse beyond. My mind transported me past distant planets, crossing the remote outskirts of our solar system where comets are swinging near their aphelions, far and still far away from our own solar system, amidst stars and more stars, and then I moved even beyond the periphery of our own galaxy, away from its billions of stars and zoomed still farther passing millions of galaxies on the way. On and on I went, and I never encountered a stopping sign that said, "Here endeth Space." Tired and disappointed, I came back again to here below. I was impressed that with my mind I could travel so far in space and return home so quickly. Joubert was right when he said, "The earth is a point in space, And space is a point in the mind". Physics heightens our awareness of the world and raises our consciousness to levels and realms that normal perception of reality seldom uncovers.

It is difficult to think of a limit to space. Infinity is baffling to the mind, and here is a perplexing endlessness that strikes us as reasonable, prompting us to conclude that space is limitless. But twentieth century astronomy revealed that this is only another illusion, for space does not extend indefinitely.

If space extended without bounds and stars were distributed uniformly in the heavens, then it would follow - and this may be mathematically reasoned out - that the infinity of stars would light up the night sky as brightly as broad daylight. In earlier centuries astronomers reflected on this question and concluded that there ought to be some distance beyond which there is no universe. This is the famous Olber's paradox mentioned in astronomy texts.

Physical space extends only as far as material galaxies have gushed forth, and this surely does not seem to be *ad infinitum*. Since galaxies seem to be advancing relentlessly every which way, space too is stretching itself to ever increasing dimensions. The cosmos, our astronomers tell us, is like one gigantic balloon that is being continuously blown to larger and larger sizes. We seem to be living in an expanding universe. What irony that on our own planet we are gradually running out of space!

Space as the stage for the cosmic show

Behind the curtain's mystic fold
The glowing future lies unrolled.   - BRET HARE

The world is a long episode of events mighty and meaningless, and on every conceivable scale. Space is the arena where things happen.

Take away space and there will be no place to put things in, hence no physical universe. We cannot picture a void of higher order where even space does not exist! But the conceptual power of the human mind is fantastic, and it can, through equations and ideas, bring forth in its field of vision utter nothingness consisting of but a single dimensionless point.

The goal of physics is to discover how the world is and would be, independently of the human mind. Here is a paradox, for it is like wanting to *describe* a scene without the use of words. A scene there could be, but its description calls for language. So there is a world, and a consciousness which experiences it. Modifying Shakespeare, one might say that all space is a stage, and all matter and movement merely players and acting.

It would be hasty to conclude from this that there was no stage or show prior to the emergence of the human mind, or that there will be none when the mind melts away. The grandest show of all must and will likely go on, even if there is no terrestrial audience to observe and applaud. But such a world of mute matter *sans* measuring mind would not be pictured or expressed, conceived, experienced or explained the way it is done by earthling-physicists. It would be like encyclopedias buried under the sea.

It is a chilling thought, a world without a human mind, ticking on for eons, without receptacle for color or response to beauty, or rejoicing at fragrance, sound, or touch. We may be able to imagine a world, but not a science, without consciousness. We may find it hard to imagine a spectacle that will go on for ages in a hall where all seats are empty. Yet, such a universe could come to pass, if our understandings of matter and energy and stellar life-spans have grains of truth.

Some older views

Speak of the moderns without contempt, and of the ancients without idolatry; judge them all by their merits, and not by their age.   - LORD CHESTERFIELD

Some ancient Greek thinkers considered space in different ways. There was the infinity (*apeiron*) of Anaximander which could have also been his view of space. Then there was the void of the Pythagoreans, referred to as *kenon*. Parmenides spoke of a Non-Being, *to mi on* in his terminology, which the later Democritus saw void as where atoms swam. Plato spoke of *chora*, a space which once emerged in the grand, and then gelled into *topos*, the space we experience today. Aristotle interpreted them as global and local spaces, comparing them to a country and a region in it. He said, "Place is what is motionless: it is rather the whole river that is place, because as a whole it is motionless."

In ancient Indic thought there was *ākāsha* which was one of the five primordial elements. It was the subtlest of them all and was endowed with the property of sound which would manifest itself only here and there. *Âkâsâ* was the vast expanse, limitless and all-pervading. It was distinct from space and time. On the physical plane, it represented the dark sky above, intangible and unattainable; and at the esoteric level it represented a mystical void whose apprehension is spiritual enlightenment. To this day, in the sanctum sanctorum of the temple at Chidambaram, famous for its sublime icon of Nataraja, the Dancing Shiva, there is a sector where naught is present: that is the *ākāsha*, the subtle symbol of spiritual effulgence.

The Indic notion of the *ākāsha* had its parallels in Western thought. The ancient Greeks called it *aether*, which came into English as *ether*. In ancient Greek mythology this Aether was personified, born of Erebus and Nyx who originated from Chaos, the pre-universe. And with Gaea, Aether produced Tartarus. Like Indic *ākāsha*, the aether too stood for the blue sky which was taken as the world beyond our terrestrial abode. This notion persisted in different forms for centuries: space was for long conceived as a super-subtle substance, existing everywhere and always.

Chinese thinkers spoke of *ch'i*, the all-pervading primal principle from which everything arose. It is ever present as the life-giving entity for the universe. It is subtle, yet can manifest itself as the material world, permeating the animate and the inanimate alike. It has pathways in the human body, and it keeps the body in balance and good health.

In the early phase of the scientific revolution, Descartes mapped space on a sheet of paper to track down figures and relationships mathematically. He spoke of the three dimensions of space. He was geometrizing space, and spatializing geometry. Descartes was also bringing to the fore the *continuity* of space: an idea that is crucial in the mathematical description of space.

The concept of *absolute space* is intuitive and ancient. Even as ships sail and fish move in the vast ocean-body, one can imagine a static expansive sea of space wherein all movements occur. In his classic work that paved the way for the science of mechanics, Newton reaffirmed this idea in a famous scholium: "Absolute space, in its own nature, without regard to anything external, remains always similar and immovable." The all-pervasiveness of space led Newton to see in it the divine principle, for is not God omnipresent? Henry More explicitly noted the many parallels between space and the Supreme Being: One simple, immovable, eternal, existing by itself, and incorruptible.

Descartes transformed the passive emptiness of the

aether into an entity with mechanical properties. He imagined vortices in the aether, the sun itself being smack in the middle of a gigantic central vortex in the universe. The existence of such an immaterial substratum was taken quite seriously by physicists in their views on the propagation of light. In the 19th century it played a role in the physicist's formulation of the laws of electromagnetism. De Volson Wood expounded on the *Luminiferous Aether*, and Oliver Lodge on the *ethereal theory of electricity.*

Immanuel Kant looked upon space (and time) from another perspective. He held that space is a mere adjustment by the human mind of the sense data that are incessantly impinging upon it. For him, space was a form in which our external sense-experiences are ordered in the brain. From Kant's point of view, space (and time) have no objective existence; they are created by our sense perceptions. In other words, our impressions of space are just that: impressions. All we understand about space are relations.

Homogeneity and isotropy
Constancy lives in the realms above.
                            - SAMUEL TAYLOR COLERIDGE.
It is one thing to study the world in our vicinity and

another to make statements about the world at large. But physics is intent on formulating a universal world-picture. It is convinced that the laws of physics that we discover here must be valid anywhere and everywhere: on the moon as in a distant galaxy, millions of light years away. Physicists are persuaded [at least they assume] that the world will be the same for no matter who from no matter where.

There are local variations. Martians can see more than one moon, and creatures in a multiple star system experience permanent daylight from the thousands of luminaries surrounding them. There are galaxies here and there, strewn all over, like ink spots on a white sheet, or mini-mounds on a plain meadow. Leaving aside such clustered clumps, the large-scale features of the universe will be the same everywhere. The universe, we say, is *homogeneous*. And so is space. In other words, space is overall the same everywhere.

Consider the universe along any direction. We see some constellations in one direction, yet others along another. But the overall aspect of space is the same along all directions. It is like standing in the snow-white of Arctic wilderness or somewhere on the Saharan sand and looking every which way. It all looks the same north or south, east or west. The same is true from any spot in space. We look in every direction, and there is no significant difference in the panorama. The universe, we say, is isotropic, and so is space.

This view by which the universe preserves common features from no matter where it is observed and along whatever direction, is what physicists call the *cosmological principle*. In a homogeneous and isotropic universe, only three things are possible: the universe remains static, it expands uniformly, or it contracts uniformly.

Measure of space

Est modus in rebus
There is a measure in all things.   **- HORACE**

Measurement is the lifeblood of science in general, and especially of physics. So we introduce a standard for the measure of space. It is convenient to start with one lengths. For this we introduce the *meter* as a standard unit of length.

Lengths have been measured since the most ancient times in all cultures, using fingers, feet, or arm-lengths. In the seventeenth century Jean Picard proposed to take the length of a pendulum which swings once every second at sea level to be a standard for measuring lengths. This would be scientific: culture-independent and unrelated to the size of individuals. After the French Revolution, other suggestions were made. A committee appointed by the French National Assembly defined the *meter* as a ten millionth part of the distance from the equator to the North Pole via Paris. Later, a platinum-iridium bar on which are etched two thin scratches a meter apart at $0^{\circ}C$ was installed at in the International Bureau of Standards at Sèvres, near Paris. Copies of this have been distributed to all nations that became signatories of the international standard.

However, the current definition of the meter is more sophisticated. The meter was defined in the 11th General Conference of Weights and Measures in 1960 as follows:

The meter is the length equal to 1,650,763.73 wavelengths in vacuum of the radiation corresponding to the transition between the levels $2p^{10}$ and $5d^5$ of the krypton-80 atom.

If you are not a physicist, read this definition again. You may not understand everything it says, but it should drive home an important point. Measurement, precision, and consensus are essential in the conduct of the scientific enterprise. Science is not one person's views about how the world functions, nor hand-waving speculations and generalities. It is based on sustained observations and meticulous measurements. That is what gives science its power and its contentions assertions about the world more reliability than those made through other modes.

Space in current physics

Physical knowledge is characterized by the fact that concepts are not only defined by other concepts but also coordinated to real objects. **- HANS REICHENBACH**

The notion of absolute space may be appealing to the meditating mind, but there seems to be no practical way of detecting it. Not that this was not tried: During the last decades of the 19th century, experimental physicists contrived the most sophisticated arrangements to put into evidence a static cosmic sea in which celestial bodies voyaged, but to no avail. Physics was forced to conclude that there is no such thing as *absolute* space.

The idea of absolute lengths and time intervals may seem somewhat reasonable. But spatial specifications have meaning only when a reference system is specified, for measured things are always with reference to something. In other words, relativity is implicit in all measurements, and this includes spatial locations.

When we work out the consequences of abandoning absolutes, we are led to a concept of space (and time) that is jolting to our intuitive apprehension of perceived reality. The most famous theory of 20th century physics, namely, the (special) theory of relativity, is essentially the mathematical exploration of the key idea of the non-existence of an absolute frame of reference, and the interpretation of the resulting formulas. It was propounded in 1905, and its author, Albert Einstein, became the best known scientist of modern times.

The relativity of space implies that the physical dimensions of bodies are not intrinsically determined. When we speak of the *length of a rod*, for example, we are unconsciously meaning a length relative to a system of reference. The special theory of relativity reveals that the length of the same rod, measured by someone with respect to whom the rod is moving, will be something different. It will, in principle, appear to be shorter.

This may sound weird, and contrary to what we would expect, but physicists have amassed ample evidence for believing in this *length contraction*. In principle, if one runs *fast enough*, one can jump over any lake, not because the speed will furnish the jumper with the required initial push, but because *the span of the lake would be contracted with respect to the fast-moving runner!* However, this length contraction becomes measurably significant and consequential only when motions are very, very fast. And by *very, very, fast* we do not mean fast as a supersonic plane or rocket, or even fast as orbiting planets which zoom with speeds like a few kilometers a second. No, here we are talking about speeds of the order of a couple of hundred million kilometers a second: unimaginable speeds, almost approaching the speed of light in empty space.

Be it noted that all the atrociously absurd implications of the theory of relativity - implications which turn topsy-turvy our common-sense notions of space and time - have been amply verified by serious and sophisticated experiments.

Geometrical space: space as points
Follow the straight line, thou shalt see
The curved line ever follow three. - WILLIAM MACCALL

Any simple line, short or long, is made up of countless points. It is useful to define such a collection of points as a space. A line may be bent or unbent, so the space in question could be curved or uncurved. They are described as *Euclidean* or *non-Euclidean*.

Then again, since any point on a line can be located by specifying just one number (its distance along one or the opposite direction from a fixed point on the line) we say that this *linear space* is one dimensional. If we consider the aggregate of points on the circumference of a circle, we have another example of a one dimensional, curved (or non-Euclidean) space, except that this one has no end-points in it. This space is therefore described as *unbounded* space, unlike the short open line which is *bounded*. The line and the circumference of the circle are also *finite*, in that their measure is not limitless.

We may extend these ideas to points on a plane to generate a *two dimensional* space which again may be Euclidean or not. The surface of a table is a finite, two dimensional, bounded, Euclidean space; whereas that of a foot-ball is an example of a finite, two dimensional, unbounded, non-Euclidean space. All the points making up the body of a cube constitute a finite, bounded, three dimensional Euclidean space.

Can we envision a three dimensional curved space? Here is a challenge to the image-making prowesses of the human mind. Try as we may, we cannot *picture* a world where three dimensional physical space suffers a curvature. The reason is simple: For curved space of any dimension we need a space which is at least one dimension higher into which it can curve. That is why we need a sheet of paper to draw a curved line, and a three dimensional space (which has volume) to accommodate the surface of a ball. But where the fourth dimension into which our three dimensional space can curve? We are limited in our imagery.

Another note on technical terminology: the measurable features of short distances in a space are said to be given by its *metric*. The metric of flat (uncurved) space is said to be Euclidean. Curved spaces have other kinds of metric, depending on the nature of their curvature. A closed curved space is said to have Riemannian metric.

It turns out that time can serve as a fourth dimension in the universe. Then we can speak of a three dimensional curved space, if such exist.

Physical Space is Non-Euclidean

... this means that physical (four-dimensional) space has Riemannian metric. **- ALBERT EINSTEIN**

The vast interstellar space up there appears to be three dimensional, Euclidean, unbounded, and infinite. But 20th century physics has revealed that it is neither Euclidean nor unbounded, nor infinite for that matter. The deeper we probe into the nature of perceived reality, the stranger we find its roots to be.

The vast three dimensional space, the abode of stars and galaxies where chunks of matter are drifting silently at random: that space is non-Euclidean. Its curvature is wrought by the massive bodies it harbors. Not unlike a sheet of stretched rubber that sags downwards in the vicinity of a heavy load on it, three dimensional space suffers a kink in the neighborhood of massive stars. The immense masses of the countless galaxies cause physical space to curve into a bounded whole which is how our universe appears to be! The curvature of space at any point now becomes a measure of how much matter is present in the region.

The curvature of space which follows from Einstein's theory is not simply mathematical poetry. It is physical enough to be subjected to experimental check. For it implies that the path of light - assumed since ages to be rectilinear - would perforce be curved in the vicinity of a massive body like our sun. This would cause apparent changes in the known positions of stars when viewed during a solar eclipse. So scientific expeditions were dispatched to Brazil and elsewhere for observing a memorable solar eclipse which came to pass on May 29, 1919. It is recorded that when the confirmations of the theory were announced in the hallowed hall of the Royal Society of London, "the whole atmosphere was exactly like that of the Greek Drama... There was dramatic quality in the staging: the traditional ceremonial, and in the background the picture of Newton to remind us that the greatest of scientific generalizations was now, after more than two centuries, to receive its first modification"

Major breakthroughs in science excite its practitioners even as the public is aroused by the crowning of a monarch, the installation of a new pope, or the inauguration of a new chief of state. Science is very much a human enterprise.

Metaphysical interpretations

The principle of relativity is a rejection of materialism...

- Wildon Carr

The picture of the world as constituted by a space which is three-dimensionally curved not unlike the earth's surface with local ups and downs is in two dimensions, emerges from the esoteric equations of Riemannian geometry, even if it teases the mind at the conceptual level. This worldview evolved from the celebrated theory of gravitation propounded by the scientific genius of the twentieth century. It is as beautiful a picture of perceived reality as anything conceived by the human mind, and perhaps a little closer to the mark of what it is all about.

The theory of relativity is erected upon a critical analysis of the notions of inertial frames and on a sophisticated formulation of a space-time metric. Since all these ideas involve abstract notions and abstruse formulas, not to mention cosmological theories and galactic recessions, the concept of a non-Euclidean bounded space is not accessible to the public at large. For the uninitiated, these ideas may seem confusing, the picture nebulous, and the formula-cluttered derivations of the results beyond reach.

Not all physicists jumped on the band wagon right away. Even in the 1920s serious attempts were made to resuscitate the classical aether. For example, like the chaos of ancient Greek thought and the ch'i of Chinese intuition, Phillip Lenard tried to introduce an *Uraether* which preceded the birth of the universe. Such attempts did not carry the day.

Instead, soon after the publication and experimental confirmation of Einstein's theories, popular versions of the Einstein model began to appear all over. These stimulated a good deal of public discussion and prompted intelligent minds to extrapolate them into theological realms of their own penchant. A famous editorial in the *London Times* declared in 1919: "Observational Science has in fact led back to the purest subjective idealism." On this question, as on others, commentators began to seek concordance between the highly technical formulations of modern physics and revelations recorded in ancient texts, including (one might add) the sacred writings of Karl Marx and

Vladimir Ilyich Lenin. Einstein himself reflected that "The mystical trend of our time... is for me no more than a symptom of weakness and confusion." That may be a description of, but not a solution to, the problem.

Be that as it may, the twentieth century view of what the poet described as "the sun-swept spaces which God made", is very different from that of a domed celestial sphere with tiny holes through which we see the fiery beyond as tiny stars, or of a limitless space extending endlessly such as earlier generations imagined.

Utter emptiness: first approximation
To find the empty, vast, and wandering air...

- SHAKESPEARE (*King Richard III***)**

Since ancient times, many reflecting minds have pondered about emptiness. Democritus of Abdera (460 BCE. - 360 BCE) proposed that the world was made up of an infinite number of never-decaying atoms in a void of infinite expanse. He pictured the eternal atoms as rigid and homogeneous and indivisible. They were moving incessantly in a void which he identified with the *Non-Being* of an earlier generation, in contrast to the *Full Being* which is eternal and indestructible. Empty space was there, said Democritus, not just to contain solid matter, but also as a region for matter to move.

Aristotle spoke out forcefully against the existence of

void. He challenged the view that emptiness was necessary for motion to occur, arguing that movement was possible in a *plenum* or fullness too: do not fish swim ever so freely in an aquatic plenum? Heron of Alexandria (c. 62 CE) considered this experimentally. In a book on pneumatics, he discussed how vacuum might occur, but also how air and water rush in to occupy any vacuum., asserting *horror vacui*: nature abhors vacuum.

These Greek ideas were discussed by Arab scholars in the Middle Ages. Scholastics like St. Thomas of Aquinas and Thomas Bradwardine were also interested in the void. Gradually, the idea of total emptiness was rejected, not because the Aristotle had said so, but because it was associated with a materialistic (atomic) theory. Insubstantial emptiness became anathema.

Sone founders of modern science thought that emptiness exerted a vacuum-force, preventing its very existence. Descartes said matter is indefinitely divisible, hence it would be impossible to remove all matter from a region. Leibniz too was a plenist, rejecting the possibility of void. They all were convinced Nature will not permit any emptiness. Therefore, they felt that something substantial or subtle must permeate every nook of available space. Speculative science can lead to competing possibilities.

Creation of vacuum

Omnia habeo neque quicquam habeo
I have everything, though nothing I have. - TERENCE

When we look at a cloudless sky on a calm day, it appears to be stark empty. Nothingness seems to be in the region above. It is also so in a sealed empty glass. But we know that the immediate emptiness above is deceptive because an invisible atmosphere is there. Likewise, the bottle also contains some air.

We may use a pump to suck the air out of the bottle. We will not be a hundred percent successful in our efforts because even the most powerful machines producing ultrahigh vacuum leave behind a few thousand molecules roaming around in the emptiness, as if to pay ever so slight a homage to ancient unbelievers in the void.

In the 17th century the issue of the existence or otherwise of void was seriously translated from the speculative to the empirical plane. This was instigated by an observation by Giovanni Baliani to the effect that water in a siphon does not rise beyond a certain height. This phenomenon was explored with care by Evangelista Torricelli. In 1643. Torricelli tried to siphon heavier liquids: sea water, honey, and finally mercury. He found that these rose to even lesser heights. His famous experiment with a glass tube a meter long immersed in a tough of mercury was carried out by his friend Viviani. The experiment was ingenious in concept, simple in procedure, inexpensive in materials, insightful in interpretation, and revolutionary in consequence. It was realized that the mercury column would not fall into the trough because it was counter-balanced by the weight of the atmosphere.

So, after centuries of speculation and debates it was established that there is indeed vacuum in the physical world. This experiment created the first artificial vacuum in a little space above the mercury column!

Soon these ideas became common knowledge, and others began to experiment with and theorize on the phenomenon. All this paved the way for barometer and weather prediction, air pump and steam engine, and for the industrial revolution. It is impressive how much resulted from a little bit of nothing above a narrow tube.

Seventeenth century physics raised our consciousness to a level of awareness it had never before experienced. And much more was yet to come.

Submerged in an ocean
All the air a solemn stillness holds.- **THOMAS GRAY**

More significant than the existence of vacuum was the conclusion from Torricelli's experiment that the atmosphere has weight. But if weight it has, and the atmosphere rose indefinitely, then we all would be crushed by it. Clearly, the atmosphere must extend to only a finite height. We can check this out by doing Torricelli's experiment up on a mountain. There the column of mercury in the tube would rise to a lesser height. This is what young Pascal did in 1648, comparing mercury heights in Torricellian tubes at Clermont-Ferrand and on top of Puy de Dôme. The column was shorter higher up, proving that the atmosphere gradually attenuates. There is indeed an immense void beyond.

Common knowledge today but arrived at only through tortuous routes. The unraveling of the roots of perceived reality is a slow process, depending more on hard work, experimentation, and reasoning, and less on speculation, wordy argumentation, and sweeping statements about origins and ends. It calls for ingenuity and intelligent observation. In retrospect, it all seems so simple. A man and his brother-in-law carried troughs of mercury and thick glass tubes: one was at the bottom and the other climbed up a hill. The mercury level in the tube slipped down atop the mountain. That was all there was to it. Voilà, we came to know that the air surrounding us does not extend indefinitely!

At about the same time, Otto von Guericke - lawyer, engineer, and mayor of Magdeburg - invented an air pump. He used it to evacuate most of the air from a huge sphere made up of two tight-fitting hemispheres which could not be pulled apart even by the strengths of sturdy steeds because of the might of the atmosphere outside. Reflect on this a little: Though we move breezily through the earth's gaseous mantle, it in fact exerts a considerable pressure on us: to the tune of some fourteen and a half pounds on every square inch. This is equivalent to carrying a considerable load on our heads all the time. Perceived reality is bearable because of evolutionarily adjusted physiology.

In due course vacuum came to be used in refrigeration, light bulbs, cathode ray tubes, and in the manufacture of thin films. Many scientific experiments require high vacuum: one reason why some day they may be conducted on the moon or in space where it is easier to produce complete vacuum.

The earth is voyaging in a vacuous void, but within the airy film: its atmosphere. Transparent and unrecognized except when we reflect on it, invisible air sustains breathing beings. As Galileo said, "we live submerged at the bottom of an ocean of air." We see aquatic creatures and we can imagine plants and creatures at the bottom of the ocean. We seldom think we ourselves are under a similar sea of air. The views behind the veil are eye-opening indeed.

It has been a long and unplanned trek from Democritean speculations about empty space to the Torricellian recognition of ultimate emptiness. The leap from speculative to empirical modes of arriving at conclusions is significant.

Utterly empty space: dynamism deep down

brhac ca tad divyam acintya-rûpam sûkshamâc ca tat sûkshma-taram vibhâti: Vast, divine, of unthinkable form, subtler than the subtle, it shines forth.

- MUNDAKA UPANISHAD

Once the extent of the atmosphere was recognized, and the suction pump was put into operation, it was easy to visualize a region were not a speck of matter is present. Much of the universe came to be pictured as cold expanses where naught but inert emptiness reigns. But patient search and observation revealed that even the emptiest regions of interstellar space are not altogether devoid matter: there are at least a few atoms per cubic centimeter in interstellar space.

The theoretical void of classical physics was not the last word on empty reality. Vacuum is not total spatial nudity, stark and still. Twentieth century physics unveiled a staggering aspect of vacuum. Calculations and conceptual penetration empty substratum point to an astounding dynamism at the core of nothingness! Pure vacuum is teeming with countless concentrations of energy that incessantly come and go. Be it in the narrow margins between the thickly packed constituents of matter or in the dark expanses of intergalactic void, a ceaseless palpitation is going on: subtle entities come into existence and vanish, like flashes of fireflies, but infinitely more quickly.

There can be no matter without the incessant appearance and disappearance of unimaginably minute entities at the core of emptiness. They are responsible for a good many features of the perceived world. They have been working invisibly since the dawn of creation until a handful of probing physicists unraveled them these many eons after the world came to be. This is the palpitation of the universe, its perennial heart-beat as it were, for should these vacuum fluctuations cease, it would be the end of the universe of matter and energy such as we perceive it.

This is difficult to grasp because it implies that some

things are being generated out of nothing and gobbled up interminably, like borrowing and returning money on a piece of paper. This is in blatant defiance of common sense and Lucretius who wrote the famous line, *nothing can be made from nothing*. In fact, however, vacuum seems to be, in the phrase of the New Testament *As having nothing, and yet possessing all things*.

One may draw an analogy with the conscious mind and the subconscious. Perceived reality may be compared to the conscious state, and vacuum to the unperceived facet where the mind is apparently without conscious thought. Underneath all the noise and action that constitute everyday life, are secret turbulences that govern our behavior. The dynamism implicit in the vacuum state is somewhat like the not-directly experienced turmoil in the subconscious.

This is the picture that emerges from what is known as quantum field theory which has opened our mind's eye to the dynamic aspect of vacuum. The so-called vacuum energy, which is one of the many surprising revelations of our probes into the microcosm, is a precisely defined and calculable entity, its actual value depending on several factors, including what is known as the cosmological constant. The notion of the vacuum state in quantum field theory requires the powerful flashlight of mathematics.

As was stated earlier, thinkers in ancient China, Greece and India also reflected on emptiness. They pictured it, not as static nothingness, but as a dynamic substratum. In their view, something subtle was going on in the void. This turns out to be an idea which has interesting parallels with the findings of current physics.

Macro and micro-space

These faces in the mirrors
Are but the shadows and phantoms of myself.
- HENRY WADSWORTH LONGFELLOW

The notion of space is not uniquely determined. There are, in fact, three levels at which space becomes relevant, and in each instance the picture is slightly different.

First there is space at our level of description. This is the *classical space* where we live like the other things. Here tops spin and projectiles orbit, the space of common sense and high school physics. This space presents itself to astronomers when they compute the orbits of planets and comets.

Then there is the microcosmic space, governed by the laws of special relativity and quantum field theory. Here, strange things happen, like the fleeing emergence and vanishing of peculiar particles. This space could be the manifestation of something more complex.

Finally, there is *cosmological space*: where huge chunks of matter abound for long stretches of time. This space has measurable curvature, warps and all. Whether it is totally closed or open, we do not have enough data to say which.

The extremes of space: the small and the large

To him no high, no low, no great, no small;
He fills, he bounds, connects and equals all!

-ALEXANDER POPE

How small an object can one see? Perhaps the tenth part of a millimeter. But there are objects much smaller. Some microbes are smaller than a fraction of a micron, and viruses no bigger. They live and die, leaving progeny and permanence, indifferent to the cosmos or to quantum theory.

Even these specks of life are made up of multi-millions of molecules, so when we descend to molecular levels, we enter a whole new realm of smallness. Molecular sizes are of the order of 10^{-10}m. Atoms are smaller still, but even they are spread out in their tiny space, extending to some 10^{-12} m. The core of the atom, called the nucleus, is smaller still, its radius being of the order of 10^{-14} m. More remarkably, microcosmic entities dance impeccably to precise physical laws.

Objects we ordinarily handle range in size from a few centimeters to a few meters. We can see and admire large lakes and landscapes, mountains and meadows of a few kilometers. There are asteroids in space, mammoth rocks a few kilometers wide, and then the moon and the earth, other planets and satellites with diameters of the order of thousands of kilometers. The sun is huge compared to these, bulging to almost a million and a half kilometers, and there are stars that are thousands of times larger even than our sun. The distance of the sun from us is about 228 million kilometers. Pluto is more than three and a half billion kilometers away from the sun!

Then there are stars trillions of kilometers away; and dimensions become staggering. Our solar system is a member of the vast Milky Way which includes billions of other stars. The diameter of the Milky way is estimated to be about 10^{18} kilometers. The closest galaxy is ten times as far as the size of our galaxy. Considering all the galaxies observed and from other considerations astronomers aver that our universe extends to some 10^{23} kilometers, a number that is more easily written down than imagined.

Such is the stupendous range of sizes structuring

space: from microcosmic minuteness to intergalactic separations and the cosmic stretch. All of this is mind-boggling, but more impressive still is the human mind that has fathomed so much of the mystery.

4

TIME:
CEASELESS PROGRESSION

Time is rhythm: the insect rhythm of a warm humid night, brain ripple, breathing, the drum in my temple- these are our faithful timekeepers; and reason corrects the feverish beat.

- **VLADIMIR NABOKOV**

What is Time?

What is time? *If no one asks me, I know it; but if anyone should require me to tell him, I cannot.*

- **SAINT AUGUSTINE**

Time is one of the most insubstantial element of human consciousness that is experienced very profoundly. Each one of us tastes a slice of time and then suddenly drops out or strays away from its course. Time seems to be with us all through our waking hours, drifting silently and ceaselessly out there in the external world as well as within the very core of our being.

ROOTS OF PERCEIVED REALITY

Saint Augustine, whose famous line is quoted above, was not speaking here about God, but about Time. Countless thinkers before and since have wondered about the nature and mystery of time. From the Upanishadic seers of ancient India and Pythagoras of reflective Greece through medieval scholastics to many philosophers and scientists, countless minds have pondered the nature and mystery of Time and acquired glimpses of its essence.

At the one extreme, thinkers have wondered about the reality of time, some contending that it is a mere illusion while others have insisted that it is as much an entity in the external world as the sun and the moon which help us measure it. No matter what, time is a perennial feature of perceived reality, powerful and useful in the scientific grasp of the world.

Time has been compared to a steady stream, gliding smoothly or rushing torrentially: sometimes lingering, sometimes galloping away at undue speed. Slow or fast, in the phrase of the poet, "there is no arresting the wheel of time." Historians refer to periods as stagnant or tumultuous. Time has been called robber of our possessions, poison, dissolver and destroyer of all. Shakespeare described time as "the king of men, he's both their parent, and he is their grave..." Yet time has also been called precious, and praised as healer of heart-aches, consoler in grief. Ovid said: *temporis ars medicina fere est*: time is generally the best medicine.

We feel intuitively that time keeps the world going and makes things happen, for a world where time did not move would be static and lifeless, more still than a painted scene, more frozen than a sculpted bust.

Our minds cannot picture a moment after which there will be no time, nor one before which there was no time. Ceaseless time seems to have had no beginning. Time, we are inclined to think, is eternal. Like expansive space and never-ending numbers, time is another baffling infinity.

Time reckoning

All time.... seems to be embraced by this cyclic quality of reversible recurrence. **- ANTHONY AVENI**

We have all been touched by the rising and setting of the sun, the waxing and waning of the moon. Some of us have even observed the changing configurations of constellations during different zodiacal periods. From trees shedding leaves and flowers blossoming we conclude that all nature is struck by seasonal changes, the ambient cold or warmth clearly provoking respite or activity. But it is not as obvious that the behavior patterns of some animals are affected by solar activity and lunar phases. It is well established, for instance, that oysters, even in depths of water where there is no light, open and close, shell-dancing, as it were with lunar phases.

Time is one of the first parameters of the physical world to be measured. Time has been recorded and reckoned In all civilizations . Aside from the biorhythm of our moods, periodic changes in our environment provoke temporal units into the nature of time. It is interesting that linear time is registered with cyclic changes.

So we have the hour, inspired by the rising of a star before daybreak; the day, inspired by sunrise and the sunset; the week, resulting from the naked-eye visibility of five planets, the sun, and the moon. The month is related to the periodic reappearance of full moon; and the year is provoked by seasonal cycles. Astronomers speak of the great year which consists of 26,000 years, while the civilization of ancient India defined still larger time units, stretching to the *yuga* which is several million years.

At the other extreme, we have defined and measured mind-bogglingly small fractions of a second like picoseconds and nanosecond durations in our experiments; modern physics has tracked down elementary particles with pathetically brief lifetimes. These breakthroughs would be impossible from speculative discourses on the nature of time.

While philosophers debate about the reality of time, experimentalists, taking perceived reality as the starting point, go forward and accomplish the most fantastic things. At the same time, at the conceptual level, modern cosmologists unhesitatingly talk of Planck time, whose magnitude, though simple to write out on a piece of paper, is beyond visualization by normal human minds.

Countless measuring devices have been constructed for the measurement of time: from sundials and hour glasses to pendulum chronometers, spring watches, digital clocks and more. What is important to note is that all time measuring devices have one thing in common: *change*. One cannot measure time if there is no change.

This empirical intertwining of motion and time measurement is the origin of what is known as the *relational theory of time*, according to which the concept and reality of time are intimately related to changes in the world. Time, in this view, is merely "the order of succession of perceptions." Some philosophers have regarded time as no more than an impression created by a series of changes while others have given it a more independent status.

Homogeneity and continuity of time

As far as their quality is concerned, temporal instants are perfectly equivalent...　　　-MILIC CAPEK

As with space, an important characteristic of time is its homogeneity; i.e. all along its ceaseless flow time is uniformly the same. There is no difference, *qua* time, between an hour or an instant some eons ago and the same in our temporal locality. Using a spatial analogy, time is like an interminable line, any sector of which is identical in its essential nature to any other.

Time is like a long highway that is being continuously created whose rate of formation is the same at every inch or mile of the path, though a host of different entities may be seen all along it. Every instant of time has another whence it emerged, and yet another into which it merges.

This implies that there was no beginning, nor will there be an end to time, for terminal points are, by definition, different from all others in being without a predecessor or successor. Even in the theological framework where the universe had a moment of creation, that was a significant point *in* time, rather than the starting point *of* time. God, the everlasting, is said to exist in time.

Recall that space has local inhomogeneities, but this is not the case with physical time. However, psychological time may be experientially non-homogeneous, some durations appearing to be denser (longer) than others.

Like spatial line again, time cannot be broken down to some ultimate indivisible unit. We speak of instants and points, but they all merge into neighboring dots in a smooth and inseparable continuity. That is why we have the image of time flowing, rather than dropping like a series of pebbles. However, it is important not to take the analogy between space points on a line and time-instants too far: such identification has led to some paradoxes in the history of human thought.

Some theoretical physicists have toyed with the notion of a fundamental indivisible time interval, dubbing it the *chronon*. Aside from some neat mathematical formulations, the idea of the chronon, once held by the Stoic Chrysippus, has not led to any significant insight or verifiable result of consequence.

Absolute time

...no master clock that monitors the heartbeat of the
cosmos. **- PAUL DEVIES**

Isaac Newton spoke of an "absolute true and mathematical time," called duration, which flowed uniformly, and distinguished this from "relative, apparent and vulgar time... estimated by the motions of bodies." This is in accordance with our intuitive grasp of the world, for one is normally inclined to imagine instants of time that is universally pervasive. Right now, there is a fleeting moment in our conscious experience of the world corresponding to which, we imagine, there is a moment that ticks away at every nook and corner of the entire universe. It is as if there is a cosmic pulse, a steady stream of subtle seconds flowing imperceptibly, carrying the entire universe along a single temporal course. It seems as if there is a cosmic simultaneity.

But the notion of absolute time through which an absolute space endures is an ancient intuitive predilection, based on our grasp of perceived reality, and it is enough for everyday discourse. It has also served as a cornerstone on which three centuries classical physics rested. But, as happens sometimes in our interpretation of the world, as we delve deeper into the roots of perceived reality, we find that there is no such thing as absolute time. Upon probing analysis, strengthened by careful observations of phenomena, physicists discovered that the notion of absolute time crumbles down.

It is remarkable that so much of significance and certainty was achieved on what turned out to be an erroneous view of so fundamental an aspect of the world. It should be noted that though physics has established with conceptual clarity and experimental certainty that there is no such thing absolute time, the notion is still useful in speaking about the age of the universe. For when current cosmology proclaims that the universe is some fifteen billion years old, an unspecified absolute reference system is implied. In the 19th century, Charles Renouvier argued that there must be "absolute beginnings."

Theologically inclined writers used to be more specific about the date and time of cosmic creation. Bishop Usher asserted memorably that "The beginning of time fell on the beginning of the night which preceded the 23rd day of October, in the year 4004 B.C." Modern defenders of ancient world views have tried to explain the Book of Genesis by suggesting that if we appropriately transform reference systems, the fifteen billion years of human science could be made to correspond to a week in the scale of the divine creator.

Information transmission

Too swift arrives as tardy as too slow.

- SHAKESPEARE (Romeo and Juliet)

Our perception of reality requires sensory inputs, and light is a primary information source for the goings on in the world. This is especially so when we take note of whatever is there in the skies above. When we fix our eyes on a star at night our first impulse is to think we are seeing the star such as it is at the very moment that flashes through our consciousness. But knowing that light takes time to travel a distance, and that the distance to travel is considerable here, we know that what we see is a body such as it was a few years or a few hundred years ago. We are in fact looking into the distant past every time we cast our glance in stellar space.

What this implies is that even if there was absolute time, our knowledge and perception of events in the world depend on how far we are from the point or region where events occur, for it takes time for light to travel the intervening distance.

This insight into the nature of perceived reality would be impossible if we did not realize that light travels with a finite speed. If, as had once been imagined, light traveled with infinite speed, then there could be instantaneous transmission of information. Since nothing physical can travel with a speed greater than light, instantaneous transmission of information becomes a physical impossibility.

This is a matter of some significance in our understanding of the world. For it appears that, in principle, it will take a finite interval of time, small or large, for information to travel from one entity to another. This has always been one of the fundamental tenets of physics. But careful experiments with photons and electrons, based on the fundamental theories of 20th century physics, suggest that in certain microcosmic phenomena, an event in one point of space may affect the status at a distant point without any lapse of time! This is contrary to anything physicists have known (or believed in) thus far.

The possibility of instantaneous transmission, even between subatomic entities, opens all kinds of possibilities. Some have argued that this makes telepathy (the reading of one mind by another) more than magic-mongering on stage, for perhaps these experiments show that minds can receive and send signals in instantaneous flashes. Scientific validation for telepathy has been sought by its practitioners and subscribers for a long time now. Some regard the results of these experiments as confirming mysticism, while others, like Leon Lederman, describe such interpretations as a "flurry of flap-doodle."

And yet, serious physicists have also been tempted to come up with models and processes in terms of which instantaneous information transmission would become a possibility. One such idea was proposed by the unorthodox quantum physicist David Bohm who introduced the notion of implicate order which is best understood by means of an analogy. Consider a table standing on the floor. The universe we experience is analogous to what creatures constrained to the flat floor would feel: to them the feet of the tables are unconnected, though in fact they are, via the tabletop. So too every event in space-time is intrinsically connected to every other via dimensions beyond the modes of perceptual reality.

Relativity of time
The happier the time, the more quickly it passes.
-   **PLINY THE YOUNGER**

We have all experienced the truth in Pliny's words, but he was referring to psychological time.

It turns out that, in a different sense, physical time also passes at different rates. One of the intellectual revolutions of our century is embodied in what is known as the *special theory of relativity*. This theory, with which the name of Albert Einstein is associated, uncovered a fundamental error in the ancient view of space and time which held the two as separate and absolute. The theory of relativity has revealed an intrinsic intertwining of space and time which results in the demolition of their independent absoluteness. What this means is that an instant of time makes sense only in relation to a point in space. This gives a death-blow to the classical notion of simultaneity.

An important consequence of this is that durations of an event or the time interval between two occurrences will also depend on the space frame of reference system. This means that if two observers set their perfectly functioning watches synchronously, and one of them gets into a train that begins to move at a certain speed, then the moving observer's watch (i.e. time in the moving reference system) will be flowing at a slower rate than for the stationary observer, and vice versa. Physicists call this puzzling phenomenon *time dilation*. For this to be significant enough to be observable, however, the train should be zooming at nearly three hundred million kilometers a second which is a technological impossibility (as of now). The veracity of time dilation is, nevertheless, a verified consequence of Einstein's theory since physicists track down and measure particles that move with such uncommon speeds.

Our understanding of the link between space and time follows from the recognition that there is no such thing (in the observable sense) as absolute motion. Every motion (or rest) that we can put into evidence by any observation, simple or sophisticated, is with respect to something else. An ant may be crawling on a table at rest, but the stationary table is moving along with the earth. The earth's motion is with respect to the sun, and the sun itself is voyaging with respect to our galactic center which in turn is receding with respect to other galaxies, and so on.

The impossibility of absolute rest or motion leads to the non-existence of absolute space or time. Though intuitively we may picture the stars as moving in a stationary empty space, a crucial experiment performed in the last quarter of the 19th century, and repeated many times since, has established that there is no observational evidence whatever for such a system at absolute rest.

Space-time

... space by itself, and time by itself must sink into the shadows, while only the union of the two preserves independence.   **- HERMAN MINKOWSKI**

The intertwining of space and time leads to a new concept of the perceived universe. The basis of perceived reality is not simply space and time, but a complex of both we call *space-time*. Since both space and time have non-discrete character, we refer to the substratum of the world as the space-time continuum.

We say that space-time has a dimension of three-plus-one, to emphasize that time as a category is in essential respects different from space. The notion of space-time, introduced by Herman Minkowski, admits of an elegant mathematical formulation. To get a general idea of this, let us consider a pair of orthogonal axes in which we take the horizontal axis to represent space and the vertical one to represent time (both with respect to a specified frame of reference). Then, any point on the plane would correspond to a point in space at a specific time. Rather than points in space and instants of time, we now describe the world in terms of space-time events. Time, to use a phrase that has become popular even in common parlance, is then the fourth dimension.

The line AB on this space-time graph would correspond to a particle which was at a space point x_1 at an instant t_1 and moved to x_2 at time t_2. On the other hand, the line CD seems a physical impossibility because it means that a particle moved backwards in time. Nor would the horizontal line EF make any sense since it implies that a particle moved from one point of space to another while time stood still. A particle that stands still during an interval of time would be represented by a vertical line like GH.

How does this essential interconnection between space and time remain hidden from our normal perceptual modes? The answer to this question is to be found in the stupendous speed of light. Light travels at an incredible speed, a fantastic speed compared to anything we are familiar with. The formulas describing processes in the world when the space-time connection is considered and those when the connection is ignored, become identical when ordinary speeds are involved.

In other words, if the universe had been designed (or had evolved) with a much smaller value for light speed, or if we happened to be sentient beings buzzing around with speeds comparable to the speed of light, then right from our conscious stage in life we would have recognized the space-time nature of the physical world.

Asymmetry of time
The dark backward and abysm of time.

- **SHAKESPEARE (*The Tempest***

There is a difference between spatial extension and temporal evolution. Given a spatial line and a direction one may move forward or backward with respect to the direction. This is impossible on the temporal line. We always move from the present *into* the future (or the future is always transforming itself into the past, whichever metaphor one wishes to choose), but never is the opposite true. There is an asymmetry in the flow of time.

This has two consequences: On the one hand we cannot move back in time. Whatever has transpired has left its relic, legacies, and records. No one can bring back yesteryear, or even yester second, for as Milton wrote in his *Paradise Lost*,

> But past who can recall, or done undo?
> Not God Omnipotent.

101

In another meaning of the word *recall*, we often recall the past: that is what memory is. Here we recognize an aspect of mind: something impossible in the physical world. The scenes and events of days gone by can be brought back in an instant to the mind's eye; even ancient episodes of the distant past that we never witnessed may be brought back to vivid reality through the pages of history. "Oh, but this is mere imagery," one may object, "it is not the past that comes back, but only its visions." But then is not the present also a vision? True, the present is tangible, but not for more than an instant. Perceived reality is a fleeting transformations of the subtle into the concrete, of the insubstantial into the tangible.

Like myriad other things, sheer familiarity makes us feel that what has happened cannot *unhappen*. However, there is a problem: If we rely on the fundamental equations that articulate the laws governing the physical world, there is something superficially unacceptable in this state of affairs, for it turns out that if we reverse the direction of time in these equations, naught is affected. The most solid laws of physics, crystallized in their mathematical purity, assure us that the world can evolve along one direction just as easily as along the other!

Physicists have been grappling with this paradox in ways that are different from their predecessors. Current thinking tries to link time's asymmetry with the initial conditions of the universe. A theoretical solution to the paradox comes from the application of microcosmic physics (quantum mechanics) to the description the universe at large (cosmology).

Though wrought with conceptual complexities such as fuzzy time and time seeping into space, this theory, introduced by Stephen Hawking and James Hartle in the 1980s, is more than an intellectual *tour de force*. It is a penetrating peep into the birth-throes of the universe, through ingenious interpretations of sophisticated mathematics that calculate singularities in black holes. The essence of the theory is that time itself had its origin with the big bang, and that it was the starting direction that determined which would be the past and which the future.

Time and causality

The Universal Cause acts to one end, but acts by various laws.   **- ALEXANDER POPE**

Time going in the opposite direction is equivalent to the future occurring before the past. Thus, if you fall and hurt your knee, the reverse order in the chain of events would be your knee bleeding first and then your falling. This is obviously impossible, because the hurt knee is a consequence of the sudden fall. Falling is the cause of the hurt knee which is the effect. Two fundamental tenets of physics (indeed of all science) are (a) every effect has a cause; and (b) the cause invariably precedes the effect.

The word *precedes* drags in the notion of time. In other words, the forward passage of time is intimately related to the notion of cause and effect, or as we also say, to the principle of causality. To imagine time moving in the backward direction is equivalent to imagining that an effect occurred before the cause. It is like hearing what a person says before he or she even utters a word.

Let us look closer into something mentioned in the last section: namely, that physical laws permit equally the forward and the backward surge of time. This is apparent in the case of the pendulum: if it swings one way, it does so with equal ease the opposite way too. However, this does not seem to be happening in all cases. A child grows into a full grown person, a full grown person does not revert to a child, like the pendulum does. In the case of most phenomena changes occur only in one direction.

Mathematical analysis reveals that any source emitting electromagnetic waves must also be receiving them at the same rate. This is very perplexing because it implies that the antenna in the radio station is not only emitting waves, but also absorbing the same waves. The emerging waves are called *retarded* because they take time to reach a destination; and the incoming ones are known as *advanced* because they are arriving before they were emitted. According to mathematical reasoning both types of waves must be present.

The paradox resulting from the symmetry with respect to time in the mathematical formulation of physical laws may be understood by means of an analogy. Suppose that you are continuously throwing balls upwards. The balls keep falling down and you keep throwing them. Both processes are occurring, and we find nothing strange in it. But suppose that you keep throwing the balls, and no ball is falling back. Wouldn't that be very strange?

In the case of electromagnetic waves, though one expects this from mathematical analysis, it would be very strange if advanced waves arrived, because that would be equivalent to effects preceding causes. So, is there a mechanism that will make this impossible though it is mathematically permissible and expected?

One explanation to the paradox was provided by the Wheeler-Feynman. These physicists showed that in fact both advanced and retarded waves are always present, but that because of the masses distributed all over the universe, electromagnetic waves are reflected back from everywhere in such quantities that all advanced waves cancel out (by interference effects), hence never observed. It is as if a counter-process is continually taking place, instigated by matter all over the universe, as if to prevent the future from materializing itself prematurely. For this, there must be large amounts of matter in the universe to absorb and radiate. This leads to the intriguing conclusion that the principle of causality is a natural consequence of the enormous quantity of matter in the universe!

Precognition and Que serà serà
If you can look into the seeds of time,
And say which grain will grow and which will not....
-SHAKESPEARE (*Macbeth*)

Another consequence of time's asymmetry is that, in the phrase of a popular song, "the future's not ours to see." While we have footprints on the sands of time, there is no trace of things yet to come. To be told that one recognizes the marks (to be) left by events yet unborn is offensive to our rational modes.

While even an omnipotent being may not be able to undo the past, an omniscient being can know the future. Here we may imagine two types of futures: One which is pre-ordained, and simply remains hidden from our view, time serving as the unveiling of events already determined. From this perspective, in the words of Omar Khayyam,

> The First Dawn of Creation wrote
>
> What the Last Day of Reckoning shall read.

We are only witnessing what was pre-ordained. If some religious traditions imagine a fate-etching God to have done the job of writing the script for the universe, mechanistic physics attributes it to the initial conditions of the universe. A ball projected in air moves in accordance with the laws of gravity, but the particular path it follows depends on the speed and angle of projection. So too, the world evolves in accordance with physical laws, but the specific modes by which atoms move and interact were determined by the initial kick that each received.

This was the picture painted by physicists in the 18th and 19th centuries. Based on their understanding of physical laws cast as differential equations, they believed that the primordial push to the evolving universe determined once and for all times how every single atom and molecule would behave at every instant in the future. This implied that every aspect of the universe that is to emerge could be foreseen by a calculating super-intelligence that could track down the dynamic states of the constituents of the world. Such a view does away with time as a significant entity, for the phenomenal world is simply like the turning of the pages of a book since everything is already inscribed. Time does not play a part in the evolution of phenomena; it is a static course on which things appear to be happening. In this view, events in the world are like images on the walls of a long tunnel through which the train of consciousness is moving. Passengers in the train zoom past the images, experiencing the scenes that come our way. In the process they feel as if the scenes are changing (i.e. that time is flowing) when, in fact it is consciousness that is hurtling forward.

This static view of time is mentioned by a character in Ursula Le Guin's *The Dispossessed* who says, "...we think that time 'passes,' flows past us, but what if it is we who move forward, from past to future... ?"

In another view the future may take unplanned turns, but omniscient God can foresee what these. In the Bhagavad Gita, the divine Krishna reveals to a mortal events that were to transpire in the battlefield. An all-knowing God, by definition, knows the future too. But do humans have the ability of precognition?

In 2600 BCE Queen Nefertiti of ancient Egypt is said to have predicted that "A king shall come forth from Upper Egypt called Ameni, the son of a woman of the South." Ancient Rome had its soothsayers. The prophesies of Nostradamus have many people. All through history, knowledge of the future has ranged from the majestic splendor of religious mythology and foretelling of royal undertakings to astrology spelling out occurrences in individual lives.

Aside from prophets, physicists who can predict the precise spot where a missile will land or the precise time when a comet will reappear, and astrologers who foretell a person's life on the basis the birth date, there are some who report recognition of events whose occurrence they witness later. In various forms, prophecy, divination, psychic powers, and the like appeal to people. Some people have had foreboding dreams, and déjà-vu experiences. Most premonitions can be brushed off as cute coincidences. But one continues to wonder whether the human brain has capacities of which we are yet not fully aware

Changes and the arrow of time

Omnium rerum, heus, vicissitudo est!
Mark this, that there is change in all things. -
TERENCE

If there is no change, there can be no measure of time. The implication is profound. Contrary to general impression, it is not time that causes change, but change that causes time. Here is a dramatic instance of cause and effect being mixed up.

Heraclitus of ancient Greece is said to have declared "everything flows, nothing stays the same." This ancient wisdom has found a thousand reformulations among philosophers and poets. But it was not until the 19th century that the connection between perpetual change and the direction of time became clear.

Careful observations put into evidence two kinds of changes. Consider first the falling of a stone from a tabletop or a child drinking a glass of milk. These are examples of *irreversible* change. Irreversible changes are changes which cannot spontaneously occur in the opposite direction. Thus, we cannot expect the stone to rise automatically back to the tabletop, nor the milk to pour back into the glass from the child's body. If the transformation in an irreversible change happened

110

backwards, it would be equivalent to time going in the opposite direction. This may be illustrated by running the film of a common event in the reverse. What appears on the screen will be the equivalent of how the world would appear if time were to change direction. Thus we note that irreversible changes are what give time an *arrow:* a direction of flow.

At first blush it might be inclined to think that if any brief event is filmed and projected on a screen one can always say if the film was running in the forward or in the reverse direction. But this may not always be possible. Consider, a pendulum swinging in an evacuated class case. If this change is filmed and displayed on a screen, one cannot say in which direction the movie is running. This is an example of a *reversible* change. Reversible changes are indifferent to time's arrow.

So if the notion of time arises only in the context of change, that of the direction of time makes sense only where change is irreversible. It turns out that irreversible changes are far more likely when large numbers are in any system undergoing change. Since ordinary matter consists of very large numbers of atoms and molecules, all changes in which atomic/molecular constituents participate are

111

irreversible. Therefore, in most natural phenomena on our scale, time does flow in one direction only.

If we were to shrink down to atomic and subatomic dimensions and observe changes that individual entities undergo, where collective and mutual exchanges do not come into play, we will be observing reversible changes. In such a world time may flow either way.

Irreversibility and return: cyclic time

Everything will eventually return in the self-same numerical order, and I shall converse with you staff in hand, and you will sit as you are sitting now, and so it will be in everything else, and it reasonable to assume that time too will be the same."

- EUDEMUS OF RHODES

When irreversible changes are examined from the molecular perspective, another interesting insight emerges. This is best grasped through an analogy. Consider a deck of cards in which the component cards are in a random sequence. Let us shuffle the deck, and the order is altered. As we shuffle more and more, say a hundred time, a series of different patterns of arrangement are generated after each shuffle. We are effecting irreversible changes, because it is impossible for the same set of configurations to arise in another hundred shuffles.

But is this really so? After all, the hundred shuffles correspond to a hundred different sequential configurations of the cards. Since we have only a finite number of cards, if we imagine performing the shuffling unceasingly, in principle, these very hundred sequential configurations will be repeated more than once in one of our future exercises.

In Hindu thought there was the idea of periodic creation (by Brahma) and dissolution (by Shiva) of the cosmos: the cycle of birth and death was not just for beings here below, but for the world at large. This would take incredibly long time-spans called *yugas*. Heraclitus in ancient Greece spoke of cosmic conflagrations too, and the idea of *eternal recurrence* was given a literal interpretation by some.

Physicists call this the *ergodic* hypothesis. A rough analogy would be that if one keeps walking indefinitely in a park, sooner or later one will trace back the paths one had already been through. The term was introduced by Ludwig Boltzmann in 1884. In technical terminology the ergodic hypothesis states that every system will pass as closely as one may stipulate to every possible state. Using the

113

mathematical theory of probability it is possible to calculate after how many attempts a given sequence of configurations will be repeated. This turns out to be a tremendously large number. Yet, as Henri Bergson put it, "A group of elements which has gone through a state can... always find its way back to that state..."

If we replace the cards by atoms, molecules and other ultimate constituents of the material world, then the aspect of the universe at any instant would correspond to a specific card configuration. The processes in nature correspond to shuffling the cards. In principle, then, after a mind-boggling series of transformations, the same configuration can (and indeed must) re-emerge. What this implies is that the universe will eventually revert to any of its current states, after a sufficiently long (but inconceivably large) lapse of time. In which case time reverts as in a cycle.

While this is very possible, even necessary in principle, in practice this is beyond realistic recognition. For one thing, the time-span required for such a cosmic recurrence is unimaginable quintillions of times the age of the universe. Yet, that this is a possibility implies a radical departure from our traditional view of linear time: At the far, far distant future, time may turn back and bring the universe back to one of its way way back distant past state.

In the cosmological models that were developed during the 1920s, Aleksandr Friedman and Albert Einstein proposed a periodic model for the universe, by which the universe would expand and contract, then expand and contract again, etc. In a way such an oscillating universe could be regarded as creating cyclic time, although it is also conceivable that after each oscillation the universe would begin to function under a quite different physical laws such as we cannot even imagine now. The oscillating model eliminates the idea of beginning and end for time.

Time ranges in the universe

All is ephemeral .   **- MARCUS AURELIUS**

It takes only a few minutes to glance through the papers, an hour perhaps to sit through a show, a few days or weeks to accomplish some tasks, a month for the moon to return to its shape, a year for the season to return, and so on. Things happen and episodes last in the phenomenal world in varying durations of time. The range of time intervals, from the fleeting life-times of fundamental particles, to the immense age of the cosmos itself, with human history and life-span in between, is impressive.

We measure time in seconds or hours, in days, weeks or years, depending on the context. They are all human-made, contrived for our convenience, provoked by our experience. In terms of these, we live for a hundred years at most, while our species has been around for at least a couple of million years. But recorded human history is barely ten thousand years, and we may reflect upon the wonders and wasteful atrocities we have wrought in the relatively short opportunity with civilization!

Mammals have been around for a few million years, and our planet itself seems to have been spinning around for a few billion years. By current reckoning, the age of the cosmos is between thirteen and fourteen billion years.

All this is general knowledge. All this information - whether dependable or changeable - comes from an analysis of perceived reality with the aid of mathematics, concepts, and ingenious instruments.

Can we define a natural unit of time that is independent of human concerns and experiences? This has been done. Consider a very small distance in the physical world, say the radius of the electron as it was pictured and measured by the beginning of this century. The time light takes to traverse this standard distance is 10^{-23} (human-defined) seconds. This may be taken as the natural unit of time.

Using this unit, we live for about 10^{32} natural times units: an impressive figure, but not longer than a hundred years. The universe is about 10^{40} natural time units.

Transcending time

You are the origin of the worlds and you are Time, their destroyer. - THE MAHABHARATA

No creature or thing we know of can disappear from space, nor jump away from the temporal axis. The notion of anything beyond the touch of time is inconceivable. The universe, from minute matter to gigantic galaxies, is embraced in the arms of times.

But would it be fair to say that what cannot be accommodated in the human mind does not exist? An entity that is both particle and wave is a conceptual oxymoron, for a particle can be constrained to a nook in space, while a wave is smeared all over: yet the building bricks of the universe are corporundals: particle-waves.

If change is what engenders time, if time is a manifestation of change, then if there is something that is changeless, it must be beyond time. Now if we define, or envision the Divine as that which is immutable, as the ever-unchanging principle behind the ceaselessly changing universe, then the Divine is beyond time.

But without esoteric mysticism fundamental physics has dragged us willy-nilly to states in this very tangible universe of ours where time as an entity naturally, not mysteriously, disappears. If anything should ever fall into the depth of a black hole, technically known as a singularity, then, says a celebrated theorem of current cosmology, it would be squeezed out of the temporal domain as well. Unbelievable, inconceivable, fantastic and whatever: but this is the translation into plain English of what the mathematical telescope unveils.

So, as with everything pertaining to the beginning and end of things, as with the ultimate essence of the most common experiences, time too is passive and serving us well when we are indifferent to it, but it becomes a teaser and baffling bully when we try to probe into its secrets.

Static and dynamic aspects of the world
Time is in some sort the Space of Motion.

- ISAAC BARROW

The notions of *being* and *becoming* have been analyzed by philosophers since ancient times. From a common-sense perspective, perceived reality is characterized by *things* and *changes*. We may refer to things there is as *beings,* and the changes as *becomings*. For something to exist in the world space is essential. We may therefore look upon space as the receptacle for whatever is there in the universe. It is the static root of the perceived reality.

For something to happen (when something happens), time comes into play. If there is no flow of time, all will be frozen indefinitely, like when you press the pause button on your video remote control. For the *becoming* aspect of the world, time becomes essential. Time may therefore be considered as the receptacle for what is happening out there. It is the dynamic root of perceived reality.

It is not surprising that space and time are fundamental in our apprehension of perceived reality. At this point two questions come to mind. What exactly are *these* things that exist? And what happens when space and time jointly act on the things. The answer to the first question is *matter*; and to the next one is *motion*.

5

MATTER:

THE STUFF OF THE WORLD

Such stuff the world is made of.

- WILLIAM COWPER

What is Matter?

When Bishop Berkeley said, 'there was no matter," and proved it, - 'twas no matter what he said.

– LORD BYRON

We recognize the world primarily through tangible things. We see things, touch things, taste things, and smell things: all this constitutes much of perceived reality. This seems to be largely a material universe, consisting of a whole range of material entities, from tiny particles of dust and sand to large planets, massive stars and stupendous galaxies. Matter does seem to be the stuff the universe is made of.

The definition of matter is no easy matter. We may say that matter is a tangible entity whose existence (as a component of perceived reality) can be felt, experienced, and established in direct or indirect ways. Matter needs some space - a tiny-tiny region ever so slightly larger than a geometrical point or a larger volume: at our level at least, all matter has extension.

Ordinarily, we find matter in one of three states. Some materials are solid as rock, others like water and oil are free-flowing liquids, yet others are tenuous as air or hydrogen. There is also a fourth state of matter, called *plasma*, to which all matter is transformed when raised to incredibly high temperatures. Plasma is not very common on earth, but much of the matter in the universe is in the plasma state: that is how matter is in the inner core of hot bright stars. Stars pass much of their radiant life at enormously high temperatures.

The material universe is emptier than filled, which means that the universe happens to be material only here and there in the vastness of its sweeping expanse. In fact, the density of matter in the universe is a paltry 3×10^{-32} kilograms per cubic centimeter. To an outside observer - if ever there is one - the universe would be one vast wasteful void, with sprinklings of matter here and there, somewhat like a dozen humans trekking alone here and there on all of an earth's otherwise bleak surface. In truth, this is not a material universe at all, but a radiant one, for its entire span is perpetually bathed in vibrant radiation. Calling this a *material* universe is like calling the oceans *naval* simply because there are ships floating around here and there.

This was not always so. In the beginning, according to current cosmologists, the density of the universe was a fantastic 10^{90} kilograms per cubic centimeter.

Though the material components of the universe occupy but a minuscule region compared to its totality, they are interesting in their marvelous properties and variety, and important too since without them there would be no universe to speak of. The few droplets of matter strewn in the vast stretches of space are what give body and identity to the physical universe.

In our field of everyday experience, it is matter, matter everywhere. Quantitatively speaking, our earth is an insignificant speck in a universe much of whose matter is concentrated is stars of unimaginably larger dimensions which are considerably more mass-packed.

Matter is the most striking feature of perceived reality. It is all around us, on us, and within us too. Bereft of matter, we and the world would degenerate into insubstantial and unimaginable nothingness: a metaphysical thought with no physical correspondence.

Variety of matter
Variety is sweet in all things. -EURIPEDES

As we look around the parts of the world replete with matter, we find abundant variety: sand and stone, water and wood, mud and mica, and much more. As if nature has not done a sufficient job, human ingenuity has concocted even more: from pliable plastics and deadly dichloro-diphenyl-trichloro-ethane to many other substances in laboratories and factories. We continue to synthesize materials every waking day: to relieve pain, to cure ailments, to make better floors, to satisfy a thousand other needs and desires.

If there was no matter, there would be no world; but could not the world have been made with just one kind of matter? Perhaps, but it is not, and if it were, how dreadfully boring it would be! But initially the world had matter of one kind only: hydrogen to be exact; soon other substances were formed from it. How this happened is more thrilling than the dénouement in any detective story. Fortunately, there is an endless variety of matter, a staggering assortment of things that make *terra nostra* such as it is, for they add immeasurable charm and beauty to perceived reality.

Every sample of matter behaves differently, has its own properties. These properties change. Thus the same substance is solid ice, liquid water, or tenuous vapor, depending on its temperature. Materials may be hard or

123

soft, rough or smooth, light or heavy, conducting or not conducting heat: on and on one can go in their description. These are some of their *physical* properties.

And then there is also richness in the range of the *chemical* properties of substances: how they burn and transform, how they store up or spill out energy in the process, how they combine with other materials or remain aloof, and so on. These too have been studied and listed in countless ways. The ability and propensity of matter for chemical change is what keeps our nook in the universe picturesque, panoramic, and throbbing with life. If the planet's conditions inhibited chemical transformation, everything would be frozen stiff in a permanence that would endure forever maybe, but it would all be inert and unchanging, dismal as in the silent darkness of distant Pluto which is a lifeless dungeon as far as we can reckon.

Elements and compounds

I must not look upon any body as a true principle or element, but as yet compounded, which is not perfectly homogeneous, but is further reducible into any number of distinct substances, how small so ever.

- ROBERT BOYLE

The universe is complexity arising from simplicity.

Though at the observational level we are struck by variety and splendor in matter, as we penetrate the deep recesses of the material world, we begin to discern surprising simplicity. However, it is not a barren simplicity but a marvelous one, rich in consequences, fruitful in expressions.

The ancients had a sneaking suspicion that this was so, for many old cultures imagined primary elements from which the material world arose. One of the earliest records of the universe evolving from some primordial stuff is in the Vedas where one uses the notion of *sat* (pure being or essence) as having given rise to the world. There was a man by the name of Uddálaka Aruni who hypothesized that from a primordial principle there arose a creative energy (*tejas*) from which came water and food. Water gave rise to life and food to the mind; he tried to prove his assertion. Thales of Miletus said that everything came from water.

Recognizing the three states of matter as seen in land, sea, and air, the ancients speculated that *earth*, *water*, and *air* were the primary substances out of which everything emerged. Realizing the importance of heat for life, they added *fire* to the list. Wondering at the boundless expanse high above, some included the *sky* (or *space*) to the basic blocks making up our world.

It has been a long route, the gradual recognition of the chemical basis of ordinary matter, known to most people today. We talk with ease about oxygen and helium, of H_2O and CO_2, but as recently as 250 years ago people knew nothing of them. Only experiments, analyses, and efforts to explain things in consistent ways enabled us to become aware of the underlying essence in the variety of things.

Thanks to the work and insights of countless investigators like Robert Boyle and Antoine Lavoisier, we now know that beneath all the colorful multiplicity of things which number in the millions, there are barely less than a hundred simple substances. We call them *elements*, borrowing from ancient terminology

The first list of elements, such as we understand the term, was published by Lavoisier in 1789: the year of the French Revolution which beheaded the founder of modern chemistry. Lavoisier's list had a mere thirty three of them, and it began with light and heat. It included commonly known metals like copper, tin, and gold; the gases oxygen, hydrogen, and nitrogen; as well as mercury, sulfur and carbon. But others carried the work thus launched, and we have since then become aware of a multitude more, bearing such exotic names as osmium and lanthanum, selenium and rubidium. Not only have

126

we come to know of their existence, we have studied and exploited their properties too.

Probing deeper into the structure of matter, we have also been able to concoct new elements: i.e. elements that did not, because they could not, last for long in the physical world. These are the more than a score of transuranic elements.

The basic elements link with one another in myriad modes and produce wondrous new substances. Every bit of matter has one or more of the basic elements. Materials with more than one element are called *compounds*.

In the presence of a piece of matter, we rarely pause to consider what it is ultimately made up of. We do not think of water as made up oxygen and hydrogen, or of sugar as a combination of carbon, hydrogen, and oxygen. Nor does red ruby remind us of aluminum and chromium any more than emerald of beryllium and silicon, or diamond of carbon pure. But the splendid spectrum of color and smell, of taste and softness, is all the result of varying affiliations of various elements, often chemically combined. How the mixing of materials results in limitless variety!

As material entities we humans are constrained by physical laws, we are puny in front of Nature's majesty, flimsy in comparison to her stupendous power. Yet, in spirit and intelligence, we sometimes accomplish more, such as creating substances that never existed before. In a way, this is only an impression, for we ourselves are the products of Nature, and all that is happening is that we serve as Nature's conscious instruments in the fabrication of even more wonders in the world at large.

The structure of matter

Seeing the root of the matter is found in me.

- THE BOOK OF JOB

Let us take a small chunk of any substance and do a thought-experiment with it. Let us suppose we break it into two bits and break the smaller one again and repeat the process again and again. In practice this would soon become impossible because the little bits would be reduced to invisible specks, beyond slicing instruments, but in our minds we can carry on the process for as long as we please.

Or can we? The question is significant because on its answer will depend how we believe matter to be. Ancient thinkers gave much thought to it and they split up into two schools. There were those who imagined one can go on and on, subdividing indefinitely any material chunk, even as one can, in principle, keep chopping a geometrical line till time runs out. Others asserted we would be forced to come to an ultimate unbreakable unit with any piece of matter. These were the *plenists* and the *atomists* of by-gone ages.

The question, like all others pertaining to the nature of perceived reality, cannot be answered by speculation. Centuries of experimentation have settled the answer. Every substance has an ultimate integral unit in which it preserves its identity. In a sense, the atomists have won. But this ultimate brick of a piece of matter is not unbreakable. It can be cleaved. But if this is done, it loses its identity. Perhaps we may make an analogy with a mound of ants which can ultimately be analyzed into so many identical ants. But if you chop down one of them, it ceases to be an ant any longer.

Atoms are not very close to one another but separated by large distances compared to their size. In solids or liquids, more so in gases, atoms never touch one another like sardines in a can. Rather, they are like fish at sea, close sometimes, or far, never in contact.

So, the perceived reality of gross matter, continuous to all appearances, is an agglomeration of minute entities, like sand grains on a beach, but far too small to be discerned as such. Who would have expected that underneath the softness of silky surfaces and the smooth

129

flow of fluids there lurks a granular structure? It is as if a myriad non-touching pebbles formed together a compact whole; their coarseness camouflaged by a deceptive apparent continuity. The illusion arises because of the scale. Our own perceptions are at a far too enlarged level, the minute discontinuities are way beyond our perceptual recognition.

The structure of atoms

Atom from atom yawns as far
As moon from earth, or star from star.

-    **RALPH WALDO EMERSON**

Contrary to the etymology of their name, atoms are not unbreakable. They have structure and components. The recognition of the composite nature of atoms is another intellectual triumph of the twentieth century. For it is only then that human ingenuity penetrated the deepest core of matter, and unraveled the marvels continually occurring in the invisible substratum of perceived reality.

Atoms consist of electrical charges. They are dynamic and spectacular in how they behave. They resemble the solar system where planets orbit around a central star; for within the atom too minute electrons are whirling around massive nuclei. The simplest atom (of the most

130

common element hydrogen) consists of a light negatively charged electron orbiting around a heavier positively charged proton. In a carbon atom six electrons are circling a nucleus with six protons and six neutrons. We may exclaim, à la Blake, that we are see a world in a grain of atom!

The emptiness pervading the atomic realm is remarkable. If the atom were enlarged to a territory a few hundred miles across, the central nucleus would be like a cottage at the center while the circling electrons would be like cars moving in distant beltways. Much of the region between cottage and cars is unoccupied space, like interplanetary emptiness. If the mass concentrations within atom were forced to come into contact with one another, i.e., if the stuff in atoms is squeezed into contiguous proximity, and all the atoms in a substance were forced to touch each other, filling all available emptiness in between, then a spoonful of matter would weigh a million tons and more.

This too puzzles our intuitive grasp of the world. For it appears that when we hold a piece of matter in hand, we are touching sheer emptiness, spotted here and there with material centers. So are our own bodies, and every other piece of matter in the world: gaping emptiness, strewn with material bits like needles in a haystack.

Ultimate entities

And in the lowest deep a lower deep...opens wide....

- JOHN MILTON

If an atom itself is cuttable, so is its core. Probing into matter may be compared to the peeling of an onion, for as each layer is undressed what remains seems to have more layers still. But we will not give up until the last dot of perceived reality is spotted. Physicists gone deeper and deeper, armed with elaborate instruments and mighty mathematics, and uncovered the ultimate bricks of the material world.

Each era of physics formulates its final findings as to where complexity halts. By the close of the twentieth century physics had a picture of matter at its core, that is cogent and colorful, and claiming at least as much finality as what our predecessors claimed for theirs.

Based on whatever we know and think today the material world is constructed of three principal kinds of point-mass concentrations. These bear the names *quarks*, *leptons*, and *field particles*. In each category there are quite a few. Now contemplate this wonder of wonders! The hardy tangible stuff of the material universe emerges from infinitesimally small punctual masses, not unlike a canvas by Seurat on which tiny dabs create magnificent sceneries.

How quarks, leptons and field particles act and interact determines the nature of perceived reality. They are responsible for the way the world behaves. These most fundamental of fundamental particles are the puppeteers, as it were, for it is to them that physics traces today every known aspect of the physical world.

Thus, the astounding statement of physics make is that every bit of observed event in the phenomenal world, tides in oceans, explosions in supernovas, orbits of planets, snowflakes in winter, or whatever, every single thing and event of perceived reality can be accounted for in terms of a handful of different entities which barely occupy any space at all: that is what *point-like concentration* means!

What is ironic in human civilization is that this worldview is supposed to reflect an ultimate revelation, a profound secret, a final answer, or something of that significance; yet, like the luxurious life of jet setters, it is the talk and truth of a privileged few: a few hundred thousand maybe in a population of six plus billion. The rest of humanity may have heard of quarks or leptons in TV specials or write ups in magazines, but they care little about it because it does not touch them in any significant way.

Transformation of Matter

...the universe delights above all in changing the things that exist and making new ones of the same pattern.

- MARCUS AURELIUS

When the log in the fireplace burns, wood turns to smoke and ash. A piece of metal rusts, seedlings grow to plants, gunpowder explodes, and food is digested: these are processes in which one kind of matter changes to others, instances of chemical transformation, as we say. Substances may of course retain their species for long periods. But they also can, and often do, undergo changes in kind, as in the examples above. Many of these changes occur naturally, and a great many are also brought about by human intervention.

Chemical transformations are instigated by heat or light or electricity. Their net effect is to change matter from one kind into another. What is happening is change at a basic level, since substances are determined by the content and configuration of their atomic and molecular core. Chemical reactions imply the splitting and forming of molecules, the switching of partners by atoms to dance with newer ones.

Material transformations are occurring unceasingly in the world around us. When a piece of paper yellows under light, and acid turns to salt, when the green of the summer leaves turns to the golden glory of the fall, silent and secretive chemical reactions come into play. Chemical reactions keep a dynamic interchange among molecules. They are essential for our biological survival for millions of them are continuously at work in our bodies, breaking up and making up molecules, fortifying blood and utilizing oxygen: the throb of life depends on complex biochemistry.

Human intelligence has understood how countless reactions come about, and human ingenuity contrives chemical reactions of interest and utility. The relevant knowledge and skill sustain giant industries that serve and support a thousand needs and desires, and incidentally pollute the environment in which we live.

In the more magical phases of human history, clever men and women claimed they could change lead into silver and copper into gold. This was magic-mongering *par excellence*, based, it was claimed, on occult powers. If superficially successful (i.e. satisfactory to an eager client and it went undetected) it could make the practitioner rich and respected. But the rosy promises of an alchemy that transmuted metals vanished with the pre-scientific past, though some present day adherents to defunct views still insist this to be a possibility.

Yet, in a peculiar way, the claims of the alchemists

were not unrealizable. In our own century nuclear physicists bring about transmutation, not in conversion of base metals into noble, but in nuclear matter.

States of matter

O! That this too too solid flesh would melt,
Thaw, and resolve itself into a dew...

- **SHAKESPEARE (Hamlet)**

Water and rocks, soil and vegetation, metals, minerals and much more are splashed all over our planet. There is also the invisible layer of air that is carried along by our planet in its cruise around the sun.

All matter we know on earth is either sturdy as solid, flowing as liquid, or tenuous as gas. These are the ordinarily observed states of matter. Matter in each of these states has specific properties as to its ability to stay put where placed, to run and flow wherever it can, or expand itself into all available volume.

As we raise the temperature of a solid, it becomes tender, and eventually melts into the liquid state. When the temperature of a liquid is steadily increased, there comes a point when it begins to vaporize. The phenomenon is readily observed when ice turns to water, and water to steam.

Ultimately, the solidity or fluidity of matter reflects

how tightly bound its ultimate constituents are. If atoms and molecules are held together in tight holds, they may shiver about their fixed positions, like the branches of trees in breeze or wind, but cannot break away from their mutual hold. As we heat a solid, we are feeding in more and more energy: it is as if the breeze turns into more powerful wind, and then the strong hold is weakened to a rope-like link, with far greater freedom for molecules to drift. So we get the liquid phase. Finally, at sufficiently high temperatures, even the weak links are broken off: every molecule becomes u independent of every other, buzzing away in all directions, bouncing off here and there from the atoms and molecules it encounters until a hard wall pushes it back into the container wherein it begins to meander every which way once again.

Depending on their intrinsic properties, substances are solid, liquid or gaseous at a given temperature. Most elements are solid or gas at the ordinary terrestrial temperatures. Dark red bromine and silvery mercury are the only elements that are ordinarily liquid.

Plasma

Sometimes too hot the eye of heaven shines.
- **SHAKESPEARE (***Sonnets: I.18***)**

Think of what would happen if a gas were heated to

ever increasing temperatures. At the core of matter are atoms with electrically charged nuclei around which are whirling smaller charged entities called electrons. At enormously high temperatures the atoms of the matter will be stripped of their orbiting electrons. Matter will be turned to nuclei in stark nudity, becoming an insufferably hot concentration of mass, gory like a creature that has been skinned, impossible to touch or even be placed in a container, for in its voracious heat it will vaporize all that comes to its vicinity.

Physics has uncovered that if the temperature of a substance reaches to extraordinarily heights - of the order of a few million degrees - then matter is transformed to yet another phase. We call this *plasma*. Pure plasma is unimaginably hot matter. Nature holds plasma only in the wilderness of empty space, far away from ordinary material concentrations: in all those twinkling stars, our sun included, whose temperatures are fantastically high.

Once it was believed that stars were burning gases. One would have thought that the plasma state was an exception. But no. Much of the matter in the universe - at least of the kind we have observed thus far - is more plasma than plain: stars are where the action is. Most matter is concentrated is in the core of stars. Stars are

138

massive beyond comprehension. Interstellar dust, planets and other rocky blobs are anomalies: These are cooler states of matter where, sometimes, life can evolve, and flowers can blossom.

But the scientific spirit will not be content with the mere knowledge that there is plasma out there. Why not create it right here below? We get fleeting glimpses of plasma when a lightning flashes and the northern nights illumine the sky, for these are in fact manifestations of ordinary matter turned plasma. Human ingenuity has succeeded in making plasma of the stellar variety also: for that is what obtains in the heart of a hydrogen bomb, and in laboratories that explore how one may tap nuclear fusion for human needs. They are awesome, threatening, and wrought with potential disaster, those horrible hydrogen bombs. But, in the context of physics and human ingenuity, we may look upon one of their explosions as a momentary mini-star right here on earth! Never in all cosmic history - as far as we know - has nuclear fusion occurred in a region that is not in the entrails of a star! No small achievement!

We have concocted weaker plasmas for more imminent use: these are gases from whose atoms, not all, but just a couple of electrons have been stripped. We call them ionized gases. They were used in the nineteenth century, long before they were recognized as such. In the last decades of the 20th century, they have come to play a major role in some industries: aerospace, biomedical, steel, and electronics. 240 high intensity light bulbs, each of 175-watt power have been replaced by just two sulfur plasma lamps which provide four times as much light. Precision plasma-processing is used of a new technological revolution.

Mass: measure of matter
... I were but a little happy, if I could say how much.
                    - SHAKESPEARE (Much Ado about Nothing)

A characteristic of matter is its resistance to change its state of motion. When push comes to shove, matter tends to oppose. It is as if it reacts reluctantly to any change in its state of motion. The degree of reluctance or resistance to change may be taken as a measure of the amount of matter contained in the body. We refer to this as the body's mass.

The conventional unit of mass adopted by the scientific community is the kilogram. It is officially defined as follows:

The kilogram (unit of mass) is the mass of a specific cylinder made out of an alloy of platinum and iridium, which is considered as the international prototype of the kilogram, and is maintained under the care of the International Bureau of Weights and Measures in a vault at Sèvres, France.

Every material body consists of a certain amount of matter, i.e. it has a certain mass. As we survey the things around us some, things like particles of dust or grains of sand have very little mass; while others, like huge boulders and mammoth mountains are considerably more massive. The masses of some smaller things like molecules and atoms are woefully flimsy in comparison. We may go beyond to moon and sun, and they have masses far greater than anything in our neighborhood, except for our dear earth which, after all, has a stature in the cosmic arena.

Astronomers talk about the mass of this star or that. But have you ever wondered about how one came to compute the mass of the earth or the moon or the distant sun? Human eyes have peered through tubes with lenses and mirrors, the mind has constructed concepts and theories. Equipped with these we have come estimate how massive a distant binary star system is. This is no mean achievement.

This is where the real excitement of science is, the ingenious and imaginative ways by which unreachable entities are brought within our scope, and how the not-directly perceived entities in perceived reality are tracked down. It is easy to discourse on the limitations of the scientific method or extrapolate its results into fantasy-land. But no pure speculation about the nature of things has ever led to any statement of significance on the measurable aspects of perceived reality.

Conservation of matter

It is sufficiently clear that all things are changed, and nothing really perishes, and that the sum of matter remains absolutely the same. **- FRANCIS BACON**

When a magician pulls out a rabbit from an empty hat, we feel that he has fooled us. We know that Lucretius was right when he said *Nil posse creari de nilo:* Nothing can be created out of nothing. And when the trickster makes the card disappear, this too we take to be prestidigitation, for we know that nothing can vanish into nothing. Even little children chuckle when they see such things.

But we also know that a brand new rabbit can come out of mother rabbit. And a piece of candle seems to disappear altogether, not by the waving of a magician's hand, but by slow and steady burning. In all such cases, chemical transformations have occurred.

Now there is another level in which the non-vanishing aspect of tangible matter has been confirmed: the quantitative level. Matter may change in form, but not in quantity. Thus, if we have wood and air in a sealed enclosure, and the wood is somehow lit, at the end of the process when all that is left in the container are ashes, carbon dioxide, carbon monoxide and other gases, we will find that the enclosure plus its contents weigh precisely the same after as before the burning. This is a principle of fundamental importance in our understanding of the physical world: the quantitative equivalent of the "nothing from or into nothing" principle of common sense. This is the *law of conservation of matter*. In the words of its first formulator, Antoine Laurent Lavoisier, "in every operation, there is an equal quantity of matter before and after the operation.

This truth about matter transformation could not have been uncovered before precision weighing was introduced as part of the scientific investigation of chemical reactions in the 18th century. The result did not come about by discussing the question only conceptually. For ages people had imagined, even without careful experiments, that bodies gain or lose weight as a result of chemical reactions.

But, like many scientific insights, the principle of matter conservation too had to give place to a more refined version of it. After all, a good deal of scientific progress consists in improving or replacing the world views of past generations.

Total matter in the universe

I believe there are
15,747,136,275,002,577,605,653,961,181,555,468, 044,
717,914,527,116,709,366,231, 425,076, 185,631, 031.296
protons in the universe and the same number of
electrons. **- ARTHUR STANLEY EDDINGTON**

Was the eminent astrophysicist and popularizer of science right in saying this? Of course he was, unless he was lying, since he began the statement with "I believe."

But it really does not matter if the number is precisely right. What is significant is the boldness behind it. Measuring mind, which has appeared in the stillness of eternity and is sparkling like the twinkle of a firefly in the utter darkness of space, declares it has figured out the number of particles in the whole universe! This is far more impressive than a microbe in the entrails of an elephant pronouncing on the dimensions of the beast.

This is the kind of thing astrophysicists and cosmologists have been accomplishing. They have measured the world and weighed it too. True, their estimates vary from era to era, each periodic news report modifying or discarding a figure held true for long. Even if we do not know precisely how much matter the universe holds, we do have an idea of how to track it down by our schemes and systems.

The method is simple enough to state. First, we figure out the average mass of a star, then the average number of stars in a galaxy, and then we estimate the total number of galaxies in the universe. Now multiply these three numbers, and voilà, we have the total mass of the universe! Yes, what we derive thus will only be an estimate because our averages and observed numbers are not that accurate.

The estimate serves us in two ways: First, it partially satisfies our quantitative curiosity about the world. After all, this is a prime motivation in the game of science. Second, it enables us to see if the observed data conform to our theories and models about the cosmos at large.

Current cosmology estimates the ratio of the mass of the universe to the proton mass be to be 10^{78}.

But here there has been an impasse. If the Big Bang model is right, as is believed by a great many cosmologists today, then there seems to be a disturbing divergence between what the theory says and what our estimates furnish. Indeed, the estimated mass is a paltry one percent of what one would expect from theory.

In the conventional methodology of science, if the results of observation are drastically different from a theoretical model, one replaces the theory and tries to formulate a new one to account better for observational data. But the Big Bang model is so persuasive in other respects that theoretical cosmologists will not easily give it up. Instead, they propose that perhaps there is something missing in our collected data. Perhaps some existing matter has been ignored in our census-report.

Missing mass, dark matter and macho

Thou, most awful Form!
Rises forth thy silent sea of pines,
How silently! Around thee and above
Deep as the air and dark, substantial, black, An ebon
mass: methinks thou piercest it As with a wedge.

- SAMUEL TAYLOR COLERIDGE

By the close of the 20th century, Fritz Zwicky surmised from his study of the motions of galactic clusters that the Milky Way should be far more massive than we had been led to believe by merely estimating the number of visible stars in our system. Could it be that we were too hasty in concluding that much of the matter in the physical universe is to be found in the stars? Were we right in imagining that only what was visible exists? If a simple stone lies in pitch darkness, and it does not glow, would it be visible? If tenuous gases filled interstellar space and emitted no visible rays, could not observe them. Should matter necessarily have to be bright to exist in space?

These are pertinent questions, and to say no, no, and no to each one of these is not only reasonable but promises to offer a clue to the puzzle of the missing mass. Maybe the universe is more massive than what we had thought. Maybe there is more than mere cosmic dust in the expanse of interstellar space. Maybe there are vast amounts of *dark matter* in the heavens.

But what is this dark matter we think pervades the world? Once it was believed that this was made up of the mysterious neutrinos that are known to zooming past through every region of space and through every star in the world. But this idea has now been pretty much abandoned. Could dark matter then consist of splinters from the primordial blow-up that caused the universe in the first place, messy discharges that accompanied cosmic birth? This was another idea popular at one time, but it too has lost adherents. Or is dark matter simply a grandiose collection of non-luminous rocks and planets, much like the asteroids of our own solar system, and/or sterile stellar debris, worn out remnants of pulsars and pent-up stars: a great many perhaps, but mere dead-weight in the throbbing stellar multitude? Some have suggested that dark matter could account for more than 99% of the mass of the universe! If so, we have again been proved wrong in our assessment of what kinds of matter populate our universe.

But how are we to see objects that by definition are invisible? By indirect means, of course. After all, that is how we became aware of Neptune and Pluto. Dark matter, if it existed in significant quantities, would influence galactic motions. Then too, if such great masses lie interspersed in space, their pull would be considerable even on light which would thus be deviated by what is called a *gravitational lens*. Astronomers have been scanning the skies and tracking the rotations of galaxies to detect such influences. They have been measuring with uncanny precision the orbital motions of minor galaxies that circumambulate our own. Their data suggest that our own galaxy must be at least five times more massive than what seems to be the case when only shining stars are considered! Searching for a descriptive acronym, astronomers have hit upon MACHO to describe such matter: *MAssive Compact Halo Objects*. It also conveys the dominant role it plays in directing galactic motions.

Antimatter

There is nothing more certain than that both are right, except that both are wrong. **- ROBERT LOUIS STEVENSON**

Many aspects of perceived reality strike us by symmetries. In the petals of flowers, in the shape of leaves, and in the form of majestic animals; in spatial directions and temporal progression, even in ethical principles like good and evil, and in human experiences like pain and pleasure, there are symmetries that impress the mind. Poets, artists, mathematicians and philosophers, all have described, captured, analyzed and reflected upon this ubiquitous feature in the physical world. Not surprisingly, it has also been forced into the physicist's recognition of the nature of matter.

Recall that the matter we are familiar with in the world of everyday experience is made up ultimately of atoms. The atoms, as noted earlier, consist of electrically charged sub-units: negatively charged light electrons and positively charged heavier protons. Now, we may wonder, why this asymmetry between proton and electron? Why cannot there be a positively charged light electron and a more massive negatively charged proton?

Theories of the micro-world suggest that if there is an electron in the physical world, there surely must be

149

another particle, its electrical reflection as it were, identical except for the charge: in other words, a positive electron. For the proton too and for every other fundamental particle in the universe, this statement holds. This theoretical conjecture, rather this conclusion from the mathematical exploration of the microcosm, was confirmed not long after it was derived: positive electrons were in fact spotted in what are called cosmic ray showers at high altitudes. We call such mirror reflections of ordinary matter, *antimatter*.

An atom of anti-hydrogen will consist of a negative proton with an orbiting positive electron. If such matter exists, one may envision anti-planets, anti-stars, and anti-galaxies somewhere out there: an anti-universe all by itself! But there are technical difficulties in accepting stellar and galactic globules of anti-matter.

Matter and anti-matter, like fire and water, cannot co-exist. When they encounter, they destroy each other instantaneously. Both will be transformed into a flash of insubstantial radiation. That is one reason we do not detect such anti-materials in the world around.

Then, where is one to get a grain of anti-sand, say, just to see and study? In the complex mammoth furnaces of present day physics, called particle accelerators, physicists routinely create anti-matte to know more about the nature of the material world. Bits of anti-matter come and go like fireflies but leave enough trails to study their properties.

Now we may ask, as with time, why this imbalance in the cosmos we know where positive protons and negative electrons, rather than their anti-pairs daringly dominate? Physicists say this was not always so. In the very distant epochs of cosmic infancy, when the universe was hot beyond imagination and as yet barely beginning, there were equal amounts of both. Then something happened causing a complex symmetry-breaking mechanism which resulted in more particles (such as we know) than anti-particles.

We eke out our existence in this world where electrons are negative and protons positive. But it is very likely that if the symmetry had been broken in the other's favor, we would be wondering why that turned out to be our fate. In the words of Stevenson, both worlds are right, and both are wrong, not one rather than the other.

Annihilation and creation of matter

The annihilation of matter is unthinkable for the same reason that the creation of matter is unthinkable....-

- **HERBERT SPENCER (1851)**

How naive, misguided, or downright wrong the

emphatic assertions of generations past sometimes sound from the vantage point of current knowledge! Based on the findings of the scientists of his own era and relying on intuitively suggestive views, Herbert Spencer, an eloquent spokesman for science in the 19th century, proclaimed that it was even unthinkable that matter could be either created or destroyed.

Today we know that Spencer's statement is not true. We now know that matter can be destroyed, not based on airy speculations by reflective philosophers, but from the work and minds of matter-of-fact investigators into the roots of perceived reality.

In 1905, just two years after Spencer passed away, Einstein propounded his famous theory one of whose corollaries is that matter and energy are equivalent and can be inter-converted. What this means is that firm and tangible matter can be annihilated, literally blown out of existence from this world, with the consequent production of an equivalent amount of energy. And out of intangible energy, tangible matter can arise.

The famous relationship $E = mc^2$ embodies this result. It is not just a formula with the name of Einstein's name. It is a powerful pronouncement of a basic fact about physical reality. It plays a role in the core of stars that shine in the universe, it is at work in the nuclear generation of electricity in our reactors, and it finds expression in the awesome blasts of nuclear explosions.

In the accelerators of high energy physics countless elementary particles) are being continuously created. If it adds to our pride, let us note that it is not simply in bringing back a dying man to life and health that Man plays God: it is also when he creates matter out of apparent nothingness, or crushes solid stuff into ethereal radiation that frail humans try to imitate their Creator.

Matter and life

Behold how great a matter a little fire kindleth!

- **NEW TESTAMENT**

It is a continuing controversy: Is there more to life than mere matter? Are creatures simply automata, robots running around, powered by chemical, instead of electric, batteries? Is man a mere machine, his brain secreting thoughts as his liver secretes bile? ? Is life just a system of bio-molecules, functioning per the laws of biochemistry, more complex than the most sophisticated gadgets, but not much different?

The debate dates back to the dawn of philosophical arguments. We all know the material dimension of life, but not all agree on its non-material. It is easy to define the characteristics of life and decide the difference between life forms and machines on that basis, but this solution becomes fuzzy at the lowest rungs of life, and at the highest levels of machines.

A significant step to the age-old question came in the 19th century when Friedrich Wöhler, the chemist, synthesized urea, an organic compound, from ammonium cyanate, a laboratory chemical. He wrote unabashedly to fellow chemist Berzelius: "I can no longer, as it were, hold back my chemical urine; and I have let out that I can make urea without needing a kidney, whether of man or dog."

The rest is history. Since then, more and more of the materials normally secreted in living organisms have been routinely synthesized in bottles and beakers. As Victor Weisskopf put it succinctly, "Chemical analysis has shown beyond a shadow of doubt that living objects contain the same kinds of atoms as non-living things."

Beyond that, probing through the spyglass of concepts and data science has come up with its own version of the genesis of life from brute matter which goes somewhat as follows.

Known as *chemical evolution*, this scheme rests on the principle that many fundamental attributes of life may be tracked down to the properties of complex chemical structures, biochemical molecules, and on the fact that under appropriate conditions some of these molecules may be synthesized in nature or in the laboratory.

The two most important types of such fundamental molecules are proteins and nucleic acids. They are large molecules at the atomic level. They result from chain-like combinations of several smaller molecules which resemble one another. Now, how did the first proteins and nucleic acids come about?

In the remote past, more than three billion years ago, and barely a billion years after the formation of our planet, there were lands barren and waste, volcanoes steaming and puffing sulfuric fumes, and oceans of salt-free waters. The earth's atmosphere consisted largely of hydrogen, ammonia, methane, and a few other gases. Gigantic clouds and torrential rains rose and fell, seeping salts from land to pristine sea. In the mammoth laboratories of the earth's oceans and airs, kindled by heat and lightning, by radiations from the sun and by other excitants, the turbulent chemistry of the early molecules churned out the first organic structures.

Carbohydrates and amino acids were thus concocted. These increased in complexity as further reactions took place. The waters of the period constituted what is described as a *primordial soup* in which mutual interactions of the components gave rise to molecules of ever increasing size and intricacy. Energy trapping mechanisms came into play. After a myriad patters and permutations, mysterious entities with the property of self-replication emerged. These grew in numbers and variety, until at last nucleic acids and proteins were formed. The miracle of life had begun.

Was this part of a Divine Plan, or did it occur by sheer chance? No one knows. We only know that these were natural consequences of the physical-chemical context in which the earth was at that time. Whatever the ultimate cause, the result, *life,* was magnificent. This was only an inkling of grander glories yet to come.

Once the spark of life was lit, the self-replicating systems began to multiply in variety. Nucleic acids with the subtle coding that preserve life patterns slipped now and then. These changes in the structures were the *mutations* which may be looked upon either as responses to the unceasing turmoil's in the earth's physical-chemical features, or as alterations merely resulting from changing conditions.

The first palpitations of life began to evolve along countless directions. As ages rolled by, and grand upheavals shook the planet's crust, ever newer kinds of plants and creatures shaped themselves. Both land and sea became homes for innumerable life forms.

Amphibians, insects, reptiles, and mammals, all evolved along with a picturesque plethora of plants and trees. After well over a billion years of such experimentation, the evolving principles brought forth the product we call humans. This conscious life form probably emerged three to four million years ago, a late comer in the series. Its potentials remained latent for millions of years. Even now, they are by no means fully expressed.

To form some idea of the mind-boggling time scales involved in all of this, one sometimes considers them in a more familiar time-reference system. Suppose that the earth was formed a hundred years ago (which we shall take to represent four and a half billion years). Then humanoids began to emerge barely three weeks ago, and the Christian era is only some twenty minutes old. Astrophysicists assure us (on this scale) that in another hundred or so years the sun will extinguish itself, after an orgy of conflagration during which it would swallow up Mercury and Venus, perhaps even our dear Earth.

Nature has taunted humans by making life too short to be taken seriously, and yet too long not to be taken seriously. That is why we continue with our plans and projects, quarrels and ambitions, in dead earnest.

Does this mean that there is no difference between brute matter and throbbing life? It is true that poems and novels contain the same words as in dictionaries, but does it follow that there is no difference between a sonnet and the accompanying word-list? Is there no difference between a giggling child and the atoms and molecules of which its muscles are formed? No one can deny there are differences, but we do not yet know their bases in the framework of physics and chemistry.

MATTER AND MIND

A man's body and his mind... are exactly like a jerkin and a jerkin's lining: rumple the one, and you rumple the other. **- LAURENCE STERNE**

What is this fleeting entity in the human body that inquires and analyzes, reflects and reasons, comprehends, calculates and creates? What is this mind that is at the root of philosophy, literature, religion and science? Does it arise from the structures that grid the brain? Is it just very complex physics and chemistry? Is it a macro-property of molecular vibrations?

Poets and philosophers speak about the powers of the mind. The Roman Manlius declared majestically: "Nothing can withstand the powers of the mind. Barriers, enormous masses of matter, the remotest recesses are conquered. All things succumb. The very heaven itself is laid open."

Every accomplishment of the human spirit has involved the mind. Illnesses have been controlled and cured by the powers of the mind. Tales, ancient and modern, have painted mind-power over matter (brain) power. Some believe that there can be mind without body. In some mythologies mind can leave the body, travel far and wide, and come back like a homing pigeon. In others, it can suck in information about events occurring in faraway places. Some argue that the mind is an open system and can work more wonders than it already does. But based on what is normally observed it is Man who goes to the mountain rather than the other way around.

We can throw a monkey-wrench in a mind by polluting the brain. A modicum of mescaline will do the job. When disease invades the brain, or brain cells age, mind withers too. Ruin the matter composing the brain, and off goes the mind. All the talk of mind over matter is true only up to a point. One is obliged to concede that mind is subservient to matter. Ultimately, matter triumphs, at least on our scale.

All this does not negate the fact that more marvelous than routine life-throb is the human mind: a flicker perhaps in the cosmic sea, but a mysterious light it is that shines brighter than any galaxy, for, but for mind, all the grandeur and glory of the world would remain unsought, inexperienced, and unsung. Matter is more powerful, but mind is more meaningful. As Robert Southwell wrote:

> Man's mind's a mirror of heavenly sights,
> A brief wherein all marvels summed lie,
> Of fairest forms and sweetest shapes the store,
> Most graceful all, yet thought may grace them more.

6

MOTION: RESTLESS MATTER

Le plus grand Phénomène de la Nature, the plus merveilleux, est le Mouvement: The greatest phenomenon in Nature, the most marvelous, is motion.
PIERRE MOREAU DE MAUPERTUIS.

Motion, motion, everywhere!
Now here, now there, now to, now fro,
Now up, now down, the world goeth so,
And ever hath done and ever shall. **- JOHN GOWER**

Movement is the most ubiquitous aspect of the world The tree branch is gently heaving at the soft touch of the breeze, the cat is running across the street, raindrops fall are falling from the clouds, automobiles speed through highways, pendulums swing routinely, the sun rises and sets: These are some instances of motion all around me. Even when I sit still for my morning meditation, I feel the rhythmic beat of my heart and the in-and-out motion of air through my nostrils.

I cast my eye on the massive rock that seems to be staying put amidst the flowers in my garden. It has been sitting there for more than a decade now, sturdy and unmoving. But no, it is not all that stationary when we realize, as noted earlier, that being part of this earth, it is spinning around with the planet, swinging through cold space at the earth's orbital speed of some eighteen and a half miles a second.

In the night sky I see countless specks of twinkling stars. They too change positions as the hours pass by. Aside from their patterned revolutions resulting from the earth's spin, they are known to be zooming at unimaginable speeds in what seems to be pitch dark space. Even the apparently fixed Polaris which guide mariners, and to which (as per Shakespeare) Julius Caesar compared himself when he said,

> I am constant as the northern star
>
> of whose true fix'd state and resting quality
>
> There is no fellow in the firmament

is not all that fixed. It is pulsating, and it is moving too.

The inert stone, motionless on soft soil, is, like every chunk of matter, made up of unimaginable numbers of imperceptible molecules and atoms. These are known to be vibrating vigorously, not staying immobile like their totality seems to be. The atoms consist of perpetually orbiting electrons that play a role in the construction of all matter

V. V. RAMAN

There is nothing in the universe that is utterly motionless. Every speck and star are engaged in a ceaseless dance, as it were, displacing itself from point to point. Hence Descartes' pithy description of the world as *matter in motion*. Let us think about this for a moment: all the range and splendor of the phenomenal world reduced to a phrase! That is part of the quest for the roots of perceived reality: to seek basic principles to encapsulate multiplicity; in formulas, when possible.

But what does all this mean? Why this restless motion everywhere, whether ordered or at random, from micro to macrocosmic entities? This no science can tell, neither physics nor formulas will bring forth meaning and purpose in our understanding of the physical world. We need poets and philosophers, religion and reflection to give flesh and form to the skeleton of scientific facts and figures.   Inspired by a view of a castle in a storm the poet Wordsworth wrote:

>No motion but the moving tide, a breeze,
>
>Or merely silent Nature's breathing life.

All motion may be viewed as the life-throb of the cosmos. When everything comes to absolute stand still, it will spell the cold silence of cosmic demise. There cannot be a universe without any motion.

Motion's variety

Diversité, c'est ma devise.

Variety is my motto.   **- LA FONTAINE**

The slow pace of the snail is very different from the fast flight of an arrow. The fall of a meteor is not the same as the flight of a bird. Clouds move and planets move, and so does waters in rivers, but not all in the same way. The pendulum moves and the ship on sea, but each very differently. And one can go on and on listing movements in the world. There is indeed a rich variety in the motions we observe.

We may describe motion by its magnitude: some bodies are moving fast, and some slow, though fast and slow are relative terms. We may also refer to motion in terms of where a body is heading towards the north or the southeast, upwards of downwards, or whatever. At any instant of time, every moving body is in a state *of motion* which specifies its *slowness* or *fastness* as well as its direction.

All the variety in motion can be put under two broad classes: motion that does not change (uniform) and motion that does (non-uniform): A body may move without changing speed and direction, or one or both may change.

Thus in one conceptual stroke we are reduce the

stupendous variety into just two simple categories.

> Things are either standing still,
>
> Or moving fast or slow.
>
> Moving things sometimes will
>
> A change in motion show.

Here we see the capacity of the human mind to classify even amid a mound of variety. Classification enables us to uncover patterns and discover causes. Careful observations first, reflections on them next. Combine the two, and many aspects of the roots of perceived reality begin to emerge. We uncover the underlying simplicity behind the complexity: unsuspected order behind confusing chaos, and patterns beneath what struck us first as pure randomness.

Thus we find only two motion-modes: uniform and non-uniform! Something profound must be responsible for this. .

Measuring motion

Now tread we a measure!' said young Lochinvar.

> **- SIR WALTER SCOTT**

To clarify the picture a little more precisely, it will be helpful to introduce some technical terms.

We measure motion in terms of the distance traveled in a given amount of time. This is what we call the *speed* of a body. When we also specify the direction of motion, we speak of the *velocity* of the body. Thus, the speed of a car may be 60 miles per hour, and its velocity could be 60 miles per hour towards the West.

When motion remains unaffected, i.e. when something is moving with the same speed along the same direction, we describe it as *uniform*. A body in apparent rest may be looked upon as a special instance of uniform motion. Driving on the highway along the same direction and with the same speed would constitute uniform motion. Thus uniform motion is any motion along a straight line with unchanging speed.

We rarely encounter uniform motion for long. Every moving thing changes its magnitude or direction of motion sooner or later. Not just vehicles and machines, people and animals, birds and insects, but falling stones, flowing waters, and blowing winds: everything seems to slow down or pick up speed, change direction, or stop.

We call motion *non-uniform* or *accelerated* when there is change in either speed or direction of a moving body. In common parlance acceleration implies only increase in speed. But in technical physics it means increase or decrease in speed (the latter is also called deceleration); or simply direction change in motion. Thus, a stone whirled around a circular orbit by means of a string is displaying accelerated motion even if its speed remains the same, because the direction of its motion is constantly changing. The moon going around the earth is accelerating too.

Speeds in the universe
There is more to life than simply increasing its speed.
– Mahatma Gandhi

While driving on the highway, a speed sign said we were not supposed to exceed the 65 mph limit. Is this fast or is this slow? Fast and slow have not absolute meanings, much less even than beautiful and ugly. The snail run over by a tortoise may complain that a speeding body stepped over it.

We can *compare* speeds. The train is moving faster than the trolley, and the buggy faster than the bug.

There is a wide range of speeds in the world. Little creatures crawl at less than a centimeter a second; a fast walker can cover a few miles in an hour; jaguars dart at sixty miles an hour. Racing cars zoom at a few hundred, while jet planes travel at six hundred. Sound in air travels with far greater speed, covering 1100 feet a second. Molecules in air move much faster: at more than 3000 miles an hour. Some stars rush through space at more than several thousand miles an hour, while grand galaxies consisting of billions of stars are fleeing from each other at speeds of the order of 100 plus km a second. Electrons in atoms whirl around the core at more than 2 million feet a second! If we wonder But no physical entity can move, even in principle, with a speed exceeding that of light: about 186,000 miles a second.

This is a world of restless bodies, from tiny electrons and mammoth galaxies, all constantly on the move with speeds spanning a remarkable range, and along so many directions.

It is even more remarkable that these speeds have been tracked down. It is difficult enough to know that sound and light travel with finite speeds, to know of hydrogen atoms and molecular motions, and become aware of electrons and galaxies. But computing their speeds is far more impressive. These are intellectual conquests in the face of which assertions about the limits of science and philosophical indictments on reductionism pale into insignificance.

Why all this restless dynamism? Why never a moment of tranquil non-movement? What if everything should suddenly come to a stand-still? No such instant can come to pass in the universe, just as long as civilization lasts there will never be a moment without a turning wheel.

For one thing, the very existence of matter depends on the motion of electrons around atomic nuclei, even as the stability of planetary systems depends on the orbital motion of planets. Contrary to appearances, stars cannot stay put in the heavens. Ours is not a static world like some ancient dungeon where lifeless articles seem to lie immobile indefinitely. stay put in the heavens.

Underlying all mute matter there seems to be some restless cosmic energy that keeps every bit of matter moving and moving for evermore. If matter is the static aspect of the world, motion is its dynamic aspect. There cannot be matter without motion, and there cannot be motion without matter.

This is one of the many dualities that make the world the way it is.

7

FORCE:
CAUSE OF UNSTEADY MOTION

Fit via vi: Force finds a way. **Virgil**

What is Force?

They seem like puppets led about by wires.

- CHARLES CHURCHILL

When we considered the variety of motions, we did not inquire about what caused such variety? One many wonder why there are these two types of motion: uniform and non-uniform? Why do some bodies remain unchanging in motion, altering neither speed nor direction, while others undergo changes now and again?

In simple cases this is easy to see: The box on the floor can be made to change its state of motion by giving it a push, the cart by pulling it. We usually affect the motions of objects by imposing direct or indirect pushes and pulls on them. Pushes and pulls are collectively known as forces.

Except for our own muscular experience of exertion, forces are invisible. And by extension, whenever we

170

observe any acceleration (change in the state of motion), we say that some force has come into play. Thus we may look upon forces as hypothetical entities subsisting in the roots of PR, in terms of which we are able to describe and explain non uniform motion.

What this means is that every time we observe non-uniform motion we say (in the world view of physics) that some force has come into play. Force, in other words, is the cause of acceleration. It is important to realize that the converse is not true: if there is uniform motion it does not mean that there is no force at all. You may push a box in one direction and a friend in the opposite direction, keeping the box stationary. In other words, forces may balance out and cancel each other's effect.

The philosopher Hume said that a miracle is "a transgression of a law of nature.... by the interposition of some invisible agent." In fact, the laws of nature themselves, rather than their transgression, are due to invisible agents. Forces are invisible agents: only their effects are observed. The wonder is that they can be precisely measured!

There are many cases of non-uniform motion: not just planets in orbits, cars acceleration and decelerating, people running and resting. But there are not too many different kinds of forces.

Measure of force: the newton

Plus potest qui plus valet: The stronger always succeeds.
- PLAUTUS

When two forces come into play, as in a tug of war, the stronger overcomes the effect of the weaker. What we call strength and weakness are in effect measures of a force.

Give a greater push to the hockey puck, and the faster it takes off. Stronger forces cause greater acceleration.. Also, the larger the body on which the force is applied, the smaller the acceleration: the toy cart is pulled more easily than the truck by the same exertion. These two facts of observation may be combined and formulated as a simple formula. If one is unfamiliar with the abstract language of mathematics, it will not hurt to see this written down: $\mathbf{F = ma}$.

Never in the history of thought were three letters of the alphabet so powerfully combined, for it turns out that this is one of the most sweeping and fruitful formulas in all of physics. Known in the world of physics as Newton's second law of motion, it couches in it every instance, actual and conceivable, of motion that occurs in the universe. That such wondrous variety can be encapsulated in such beautiful brevity is a marvel. With the aid of this simple-looking mantra of modern science we have been able to explain and understand much of the complexity in the world, when we know the forces involved in each situation.

Using the formula given above, one defines in the world of physics the intensity (or strength) of a force in precise quantitative terms. In the course of the 19th century a convention developed by which units of physical quantities would be named after eminent contributors to the field, or, to put it in the more pungent language of Hogben, the fashion was set "for combing obituary notices of departed worthies to upholster names of physical units." So the unit of force was named *newton*: note that one does not use the capital letter in writing out the unit, as that would refer to a person. However, the abbreviation for the physical unit is given (by convention) by using capitals, like N for newton.

Source and target: third law of motion
Cause and effect are two sides of one fact.
- RAPLH WALDO EMERSON

You push the book on the table: the push is the force; you are the *pusher*, the *source* of the force; and the book is the *pushee* or the *target* of the force. The child pulls the string: the pulling is the force, the child is the source, and the string is the target. In every instance we can distinguish between these three entities: a force, its source, and its target.

Observations reveal a most interesting symmetry in the roots of PR: whenever a force (which invariably has a source and a target) comes into play, another force of equal magnitude appears also in which the source and target are the reverse of the first force. In other words, here the source of the first force becomes the target and its target becomes the source. Thus, when you push the book with your hand exerting a certain force, the book pushes back your hand with an equal amount of force.

The existence and equality of action-reaction pairs of forces in known as (Newton's) third law of motion. We usually refer to one of the forces as *action*, and the other as the *reaction*. Like cousins, action and reaction are always reciprocal. If we consider one as the action, the other becomes the reaction and vice versa.

In the 18th century a paradox : When a horse (source) pulls (action) a carriage (target), then the carriage (source) must pull (reaction) the horse (target) with an equal force. If that is the case, one argued, then how come the carriage is moving? The answer lies in the fact that horse and carriage do not move with respect to each other, but with respect to the road. All motion on surface occurs from the operation of the principle of the action-reaction principle. When we walk, we kick the road back, and the road kicks us forward. That is what wheels do too. For ages human beings have walked without recognizing this root of PR!

Another paradox in this context is how any change in motion is possible at all if all forces come in equal and opposite pairs. Should they not cancel one another out, leaving no net force to cause changes? It is important to remember that *the action force and the reaction force act on different bodies, never on the same body*. That is the reason changes in motion of different bodies is possible.

Weight

Weary the cloud falleth out of the sky,
Dreary the leafy lieth low.
All things must come to the earth by and by.
Out of which all things grow.   **- OWEN MEREDITH**

We live in an age when people are conscious of their weights. Weight watching has become a national concern, and has given rise to health foods, obsession for low calorie foods, attempts at fasting, jogging, gyms, etc. But not one in a hundred may be able to tell what exactly one means by weight. People may laugh if told that by merely flying to the top of a mountain one can lose some weight.

But this is true, because weight (on earth) is simply the force with which the earth draws us towards its center, and the farther we are from that center the less is that force. We note right away that being a force we must measure weights in newtons and not (as they do in doctors' offices) in kilograms. It's okay to measure weights in pounds because the pound is another unit of force. The "overweight" person is trying to lose *mass*, not weight.

Everybody in the vicinity of the earth experiences this force called weight. And if one is sufficiently far away from the earth, this force becomes negligible. When one is close enough to any other celestial body (the moon, another planet, the sun, etc.) one will experience similar pulls. Depending on the mass of the pulling body our weights would be less or greater. On the moon we would be much lighter than here below, but if transported to Jupiter we will be glued tight to the ground. Weighing incredible tons, it will be impossible for us to even stand up.

It is weight that causes things to fall to the ground; but for weight things would hang in thin air. It is weight therefore that keeps us here on earth; without it we would be flying all over space. If we are glued to our earth it is because of the force of weight. It enchains us to this speck in the universe, yet it provides us with the security of a home in the cosmos. Again, weight is also the force that holds on the mantle of air that is hugging the planet. If the earth's hold were too slight (as in the case of the moon), all air will fly away and there will be no atmosphere. Indeed, as it is many gases are escaping from our atmosphere continuously.

Friction

It boots not to resist both wind and tide.

- SHAKESPEARE (III Henry VI)

When an object given a push on a tabletop, it soon comes to a stop. What made it motion change? When we exert a force on a heavy box on the floor, it does not easily move. When a wheel is spun on an axle, it eventually comes to a stop. Whatever caused this change in motion? Why is there no change in motion even when a apply a force?

All such phenomena are explained when we recognize the existence of another kind of force called (the force of) *friction*. Friction is a commonly occurring force whose principal property is to prevent or stop motions on surfaces. With solids, friction prevents the start of motion itself, whereas with liquids and solids it comes into play only after motion has begun. In all instances it is because of friction that moving bodies slow down and halt when they are in contact with surfaces, and turning wheels rotate slower and stop. Friction causes the wear and tear of materials. It is to minimize these that one uses roller bearings and lubricants.

Friction results in the dissipation of energy as heat. In other words when friction slows down a moving body, the body loses some energy. We shall discuss more of this later.

Though friction seems to be a nuisance in the context of motion, ironically friction is also helpful, indeed indispensable, for locomotion. If frictional forces do not come into play between feet and ground, we cannot walk. That is why we slip on ice and oily grounds. One imagines that it is the turning of the wheels that causes a car to move. Rather, it is friction that makes this possible: that is why the car will not move with bald tires on a snowy road. We cannot lift an object with our fingers nor play the violin or the cello, were it not for frictional forces.

In mid 19th century Charles-Eugène Delaunay calculated with the utmost precision motion of moon and found some minor discrepancy between his results and the data of observation. He suggested one possible cause of this: the slowing down of the earth's rotation as a result of the friction arising from the motion of the tides ! The vast masses of ocean waters are in continuous motion through waves and tides. They rub against the ocean floors and splash on the coastal surfaces. All this means friction and dissipation of energy. The net effect of it is to slow down the earth's rotation. This would cause a lengthening of the day. It turns out that this loss is negligibly small in the span of human history: barely 0.016 s in a thousand years. It is not the amount that is significant in a case like this, but our awareness of a process like this. Who could have thought, without careful observation, precise calculation, and ingenious concepts, of the earth slowing down in the course of thousands of years, and days getting longer as a result?

Although the bare effects of friction were known since a very long time, it was thanks to the detailed experimental studies of Coulomb in the course of the 18th century that we have come to know a good deal about various aspects of friction as a commonly occurring phenomenon. He is credited with the founding of the science of friction. His contributions to the science of friction were exceptionally great. "Without exaggeration," noted two scholars, "one can say that he (Coulomb) created this science ".

Tension

Skilled to pull wires, he baffles Nature's hope,
Who sure intended him to pull a rope.   **- J. R. LOWELL**

The chandelier is hanging from the ceiling, attached to a chain. I know the earth is pulling the lamp through the force of weight. But there is no change in the state of motion (rest) of the lamp. I have to conclude there is an equal and opposite force pulling the lamp in the upward direction. This force must be exerted by the chain to which the lamp is attached. We call this the *tension* (force) exerted by the chain. This is another type or ordinarily occurring forces on our scale.

Any rope or string, chain or rubber band, exerts tension when it is attached to an object and stretched. In our technological world, therefore, there are a great many contexts in which they appear. In belts connecting wheel, in suspension bridges and in lamps from ceilings, in transmission wires and in elevators cables, in pulleys and cranes and children's jump-ropes, we find countless instances of tensions.

While weights and fractions occur more generally, we do not observe tension forces in natural phenomena, for there are no ropes hanging from cliffs, no wires stretched across valleys, nor strings stretched in the wilderness. In the biological world tension forces do come into play, as in the tender cobwebs and in tendons connecting joints.

When we stretch a string, there is only so much tension it can bear. Hanging a massive rock from a slender fiber is not a wise idea. Materials have a finite tensile strength, we say: if stressed beyond, they break. Even columns of liquid have tensile strengths. It is not easy, for example to break a long column of water because its tensile strength is considerable.

Many key insights of modern physics had occurred to some ancient thinkers also. A case in point is the role of the tension force in the equilibrium of bodies which was studied and analyzed in remarkable detail by Leonardo da Vinci already in the 16th century.

Normal contact force

Vi victa vis: Force overcome by force　　　- CICERO

When I am standing on the floor, I know that my weight is pulling me towards the center of the earth. But there is no change in my state of motion. I am forced to conclude that another force is balancing my weight. This is a force exerted by the floor on me. Because of my weight I am pushing the floor downwards. Because of my (source) push (action) on the floor (target) in one (downward) direction, the floor (source) is pushing (reaction) me (target) in the opposite (upward) direction. The force which a surface exerts on a body impressing upon it is called a *normal contact force*. Thus we see that normal contact forces are direct consequences of the law of action and reaction. Like all forces, these too cannot be seen save through their consequences.

Normal contact forces prevent the rude penetration of bodies into surfaces. It is because of these forces that objects can rest on tables, we can sit on chairs, and ladders can lean on walls. What this means is that such most ordinary things we do, like standing and sitting, reclining and lying on bed, are possible only because of the so-called third law. The roots of PR have the most unsuspected aspects.

Let us emphasize the difference between a force and the experience of a force. When we press against a wall, we feel the push of the wall against us. Similarly, we must distinguish our weight from the experience of weight. That experience comes only when we are standing on a surface. It is absent when we take a dive or if we happen to fall. Then we feel we are weightless. When we are on an elevator which is accelerating downwards, the floor is pressing on us with less than the force of our weight, and we feel lighter as a consequence. If the downward acceleration of the elevator is equal to that caused by the weight (the acceleration due to gravity), then we will feel no weight at all. This is precisely what happens to astronauts in earth satellites. The orbiting satellites are in free fall, and so the astronauts experience complete weightlessness.

Buoyancy

Bodies float in liquids and gases because of the force of buoyancy.

When we enter a pool, we begin to feel lighter. No we have not lost our weight, the downward pull; but an upward push by the body of water cancels out a part of the weight. This indeed is the property of all fluids: to gently push in the upward direction anything and everything that is immersed in it. We call this the *buoyant* force. But for it ships can't float, not swimmers swim, for we need forces to counterbalance the earth's pull downwards we call weight.

If a plank of wood floats and lead ball sinks, it is because the upward thrust of water cannot be of any magnitude. It depends in fact on how much a body is immersed in the water. If the buoyant force exceeds the weight of a body, then the body will not sink, but rather would be pushed to the surface. If the buoyant force is less than the weight of the body, the body will go right through and sink. If the two forces are equal, then the body can remain pretty much anywhere within the fluid.

That water exerts a buoyant force should be a matter of simple observation for anyone who was gone into a pool. The story goes that the principle of buoyancy came to the great Archimedes in a flash one day while he was in a public bath. In his great excitement, we are told, he ran stark naked into the streets screaming "Eureka! Eureka!" which in the language of the day meant, "I have found it! I have found it!" This story and cry have come to symbolize the joy of discovery.

The buoys where ships come to anchor are meant to warn the pilots of rocks, cables, and the like. These floating objects suggest that buoyancy is a property of liquids. But gases (and air) also display buoyancy. That is, , gases also exert an upward force on solids immersed in them, diminishing the weight. It is because of the buoyancy of air that balloons float in air.

Much of the continental land mass is high above the sea level, enabling us to live on ground. Here too buoyant forces are at play: The granite and basalt rocks which constitute our continents are lighter than the material of the earth's rock mantle which lies underneath. So, the lands literally float up above. This phenomenon is referred to as *isostasy*. What an unexpected revelation? Archimedes' buoyancy in the bath-tub and the drift of the continents on the earth's mantle are in fact related! Here is another root of PR that is utterly hidden from our normal view. Only observation and reflection bring it to our understanding.

Change of forms

And never did Grecian chisel trace
A nymph, a Naiad, or a Grace
Of finer form or lovelier face.   **- SIR WALTER SCOTT**

We see bodies of various forms and shapes. Why are they not all the same? And how can their forms be changed? Here again forces are involved. Forces not only cause acceleration, they can change the forms of bodies. When you press on a balloon or pinch a cheek, you are causing deformation: rather the force you are applying is causing deformation.

Deformation is possible because most solids are compressible. A body that is not so is said to be rigid. Rigid bodies suffer no change in shape even under great pressure, but there is no body in the world that is perfectly rigid. From the bouncing ball to strong steel, everything is deformable.

When you sit on a cushion or place you head on a pillow you are deforming the body because of the force you are exerting. When a child handles the play-dough, it too is deforming the substance. When the pizza-maker rounds the mass of moist flour and flattens it to shape, we see what the deforming power of force can accomplish. When an artist manipulating clay, the material is deformed too.

When we speak of evolution it is of species that we think. But matter too is evolving in shape and form, under the action of the forces. The sun and stars were not always the same, nor the earth and its mountains and valleys. Eons from now, when you and I and our species will have gone, sun and moon will all look different, if On-lookers there ever be, for the forces will never cease.

The deforming effects of forces are seen not just in the compression of balloons or strokes of the blacksmith, but in the transformation of the world at large.

Gravitation

That very law which molds a tear
And bids it trickle from its source, -
That law preserves the earth a sphere,
And guides the planets in their course.

- SAMUEL ROGERS

The effects of celestial bodies on things terrestrial had been suspected since very ancient times. Effects there sure are, but not of the kind imagined by planetary fortune tellers. Rather, they are related to a magnificent force that reigns supreme in the cosmos at large. We call it *gravitation*.

The law of universal gravitation, especially in its mathematical aspect, was the deep insight of Isaac Newton though others before him had toyed with similar ideas. The law itself is simple to state: Every mass in the universe attracts every other mass with a force that is proportional to the masses and inversely proportional to the square of the distance between them. This means that if we double or triple one the masses the force between them will also be doubled or tripled; and if we double or triple the distance between them the force will be diminished to a fourth or a ninth part. As it says nothing of the fates and fortunes of people, young or old, nor contradicts any sacred passage from a holy book, no parent has argued that gravitation is only a theory or that it ought not to be taught in schools.

Newtonian gravitation has been one of the most successful of physical theories. In one grand sweep it accounted for the Kepler's laws of planetary motion and extended them to the moon and to other satellites, as also to distant stars and planetary systems. Newton himself believed that chemical combinations are also due to gravitational attractions between the atoms of substances, because he used to term to signify any force between bodies After all, this was the only kind of universal force then known. As Mellor pointed out, "Newton applied gravitation concepts to atoms, and in this sense, he was the founder of molecular as well as of celestial mechanics." As astronomical observations extended the role of gravitation became clear in the harmony of the universe at large.

The quantitative aspects of gravitation gave rise to many challenging problems calling for sophisticated analysis and complex calculations, like lunar precession, the three body problem, the erratic orbit of Uranus, the motion of mercury's perihelion. Some of the best creative minds tackled these long before computers were conceived, when calculations meant hard work and ingenuity. The predictions from gravitation were impressive too: Halley and others could predict the return of their comets, and planets beyond Uranus could be discovered on paper.

The theory of gravitation also contributed much to mathematical thought. The grand subject of celestial mechanics rose in our efforts to explore and apply the gravitational thesis. Gravitation inspired other domains of physics as well. Priestley and Coulomb derived the laws of electrostatics by suspecting parallels with gravitation.

Today we know that the relevance of gravitation lies primarily at the astronomical level. Gravitation keeps systems stable over long stretches of time. Were it not for gravitation there cannot be orbital motion of planetary bodies and in stellar galaxies. Masses from the primordial Big Bang would have splintered helter-skelter along divergent paths, receding away independently for as long as time will tick. It is mind-boggling that Pluto at billions of miles away moves with mathematical precision in orbits and with periods dictated by the gravitational sway of the sun.

Nor would the sun and stars be possible without gravitation. For these are gradual but massive accumulations of matter, persuaded over eons to mutual adhesion, primarily through the gravitational force. It is gravitation that compressed matter of mammoth magnitudes to generate temperatures high enough to induce the nuclear fusion necessary to set the stars aglow. We often think of the sun as the source of all our energy. Maybe, but the sun could not have been formed without gravitation

Ironically it is gravitation again that ruthlessly decimates stars in their old age. With relentless pressure

Gravitation smashes matter at its skeletal void, transforming stars of ageless shining into white dwarfs and pulsars and more. When conditions are appropriate gravitation can act with frightening fury and cause a gravitational collapse resulting in the monstrosity called the black hole. In the heart of black holes space and time are mutilated, and the very laws of physics cry out for redefinitions.

Thus we see that the Newtonian picture of gravitation as attractions between massive bodies is effective and adequate in a whole variety of contexts: planetary motions, falling bodies, and stellar evolution. No wonder it ruled the realm of physics for more than two centuries, until Einstein's wand metamorphosed it alls. Now gravitation has become a kink in space-time, the result of a warping brought about by too much mass in the unfathomable fabric of space. Curved orbits are merely motions along the shortest paths in a curved space-time, even as uniform motion is merely persistence in a Euclidean manifold.

Einstein's formulation of gravitation is eminently more elegant from a mathematical stand-point. More than that, it implies phenomena like the bending of light rays in the vicinity of stars that have no analog in the older view.

The Einstein theory says there are gravitational waves which propagate through interstellar space. Gravitational waves may arise from catastrophic events, like supernova explosions, or from periodic processes, as in double or spinning stars. Theoretically, even a spinning rod should emit gravitational waves. Human ingenuity knows no bounds. We concoct ideas and contrive devices incessantly. The puniest of perturbations are detected and eventually measured. So with esoteric apparatus and painstaking dedication some have tried, even announced their instruments have responded to the faintest palpitations of gravitational waves, perhaps from the core of our galaxy, perhaps from the orbital variations of binary pulsars.

Then again, in the framework of current physics, with every fundamental force are associated certain fundamental particles called quanta. The quantum of gravitation is called the graviton. More on this later.

Gravitation is the hurdle we crossed before we learned to stand erect. It is the string that binds us to our planetary habitat in the universe. It is the music by which the planets glide in Keplerian orbits, and proximate double stars round each other, as do neighboring galaxies as well. It is also the awesome power that sets fire to stars and crushes them to smithereens eventually.

This cosmic grasp all that is matter also causes insubstantial space-time to twist and turn. Like Nature herself, gravitation is expressive in a thousand ways, active forever and everywhere, holding the world together.

With all its simplicity, gravitation is still a mystery. Who can say what possibilities are in store when we unveil her secrets even more?

Actio in distans: **Action at a distance**
Causa motus nulla esse potest in corpore nisi contiguo et moto: There can be no cause of motion in a body except from a body in contact and moved.

- THOMAS HOBBES

Try to move something without touching it directly or indirectly, and you will fail miserably. Really, there is no way one can cause any motion except by making contact with it one way or another, wither by placing our hand or foot on it, or with strings or sticks or whatever. Any other mode of causing change of motion would be action at a distance. We would describe such action as magical. Even Yuri Geller would confess it is clever illusion. It will certainly not be mechanical.

We see this happening all around us and are not in the least surprised. When the vase falls from the table,

192

or the apple from the tree, there is no contact with the descending body! We speak of gravity so glibly, of the sun whirling the planets as if it had a ropes tied to them. This is no mean puppeteering feat that the sun does, pulling bodies millions of miles across space. Truly, there is something spooky about this gravitation bit.

So when Newton audaciously proposed it, some of the best minds of the time were not exactly enthusiastic. Cartesian pundits, Huygens, Leibnitz: not one of them thought this made any sense. Voltaire informed the world that for forty years after its publication (1687), barely a score of scientists took it seriously. For the next two centuries many critical minds found it difficult to accept the idea of an influence propagated by inanimate matters through the abyss of empty space.

But once its powers were recognized, one could not brush off the idea as a crackpot attempt to bring back angels and spirits in physics. When the idea is cast in its mathematical mold and explored, you can squeeze out of it all the laws of Kepler on a clean sheet of paper. Why, if no one had ever seen a planet in high heavens, one can still say how they ought to be moving move and how fast by simply working out the consequences of Newton's laws, including gravitation. So here is this mysterious action at a distance (or *actio in distans* as they used to say in Latin) which no one likes to accept, but which, if accepted, explains so many things, including when Halley's comet will appear again.

That is, in effect, how physics works. Give me an idea from which some aspect of perceived reality can be explained, and I'll buy it, even if it seems somewhat weird to begin with. But our reluctance, indeed difficulty, in accepting action at a distance also exposes the intellectual incapacity of the scientific mind to accept something that cannot be mechanically pictured. We see this happen every day and night, all around us and in the universe at large: gravitation propagated without the benefit of a material medium. Yet, we are unable to concede that this can be so. We want something in between so that a force can be transmitted through a so-called medium. Hobbes was quite right in terms of the motions we ordinarily see in the terrestrial context, but dead wrong when it comes to gravitation.

Interactions

Born under one law, one to another bound.

- FULKE GREVILLE

The gravitational attraction between two bodies is mutual. In accordance with the third law of motion, when one body attracts another with a certain force, the other does the same. Bodies act on one another: they interact. Interactions are what bind the world as one. Forces are interactions.

They do more: They cause change. They make things happen. They produce phenomena. No interactions, no change, no phenomena. The world will remain as it ever was, drab and static, uneventful and actionless. Mere matter won't do to keep a world functioning, for what is functioning but exertion and dynamism? Gross matter in the universe without interaction would be like lone sculptures, cold and lifeless in an empty hall. What an utterly boring world that would be!

There are interactions everywhere, mutual and multitudinous. There is not a spot in the world where interactions do not come into play. They are there deep in minute microcosm, right on our levels of activity, and in the grand cosmic scale too.

There is being and there is becoming in the world. Matter is the being in our world, and interactions cause becoming. There is all there is to it: things, made up of molecules and atoms and electrons and what not; and happenings to things caused by forces, by interactions.

Forces like weight and friction, tension and normal contact reaction are what we measure and calculate at our level of experience. They are easily observable, or at least their effects. But how do they come about? They arise from more fundamental features of the world. Their roots lie deeper, beneath perceived reality. And their manifestations are many. As we noted, the weights of objects are due to gravitation: one of the fundamental interactions under girding the universe. As we have seen, aside from weights which make apples fall and stones to sink, gravitation causes planets to turn, stars to form, and keep the universe together.

And there are a few other fundamental interactions as well. Let us briefly consider them now.

Fundamental interactions

Every particle is... reacts in some way to the presence of every other one, although that reaction may be anything from the barest nod of recognition to the explosive violence of annihilation. **- KENNETH FORD**

Ultimatelythe world is made up of many *elementary particles* which are minute beyond direct perceptual recognition. Their variety is not unlike the plethora of species that compose the realm of life-forms. And even as living entities act upon one another and maintain a biosphere, the ultimate particles incessantly interact in different ways, giving rise to this marvelous world of perceived reality.

The human mind has been able to track down and categorize the number and nature of the fundamental interactions that make things happen. We have already referred to one of them: gravitation. Physics has discovered that there are only three others. What this means is that in terms of just these three four fundamental interactions we can understand every known (physical) feature of perceived reality. Yes, just four types of fundamental reactions are at work to make the universe click the way it does.

One of the other three is: *electromagnetic*. The other two are described by more prosaic, if more intuitively graspable, terms: *strong* and *weak*.

The electromagnetic interaction is due to the existence of electric charges in the universe. It is responsible for so many aspects of perceived reality, it deserves a chapter all for itself.

The terms weak and strong might suggest that not all interactions are equally potent in their effects. This is not so. The terms connote a quantitative feature that has little to do with ultimate effects. If we list fundamental interactions in terms of pure strength alone, gravitation will be right at the bottom. It is far, far weaker than the weakest of the other three. Yet, and this is the marvel, its embrace extends to distances that are infinitely greater than how far the strongest of the interactions can exert. Just consider the sway of the sun over Pluto at more than two billion miles away, or of galaxies separated by light years which are bound to one another. Who will dare to say that gravitation is weak?

How do interactions arise? What are their specific effects? How, if at all, are they interconnected? These are among the countless questions that intrigue physicists. Some answers have been found. Others are still a mystery.

8

ENERGY: ROOT OF
NATURAL PHENOMENA

Energy is Eternal Delight. – WILLIAM BLAKE

Everyday uses of the technical words

A word is not a crystal, transparent and unchanging, it is the skin of a living thought and may vary greatly in color and content according to the circumstances and time in which it is used. - **OLIVER WENDEL HOLMES**

(X) The term *energy* is used with a variety of meanings, We speak of an energetic person, we say we are devoting a lot of time and energy for a task, we refuse to walk farther because we have no more energy left, etc.

In recent decades the word has acquired new importance. It is well known that energy is something we need for our comforts and civilization, indeed even for our survival. The industrial revolution has energy at its base. All applied science is a manifold use of energy, essentially to serve human needs and wants.

It is often said that our energy resources are being depleted. Environmentalists protest the damage wrought on the world by technology and plead for a slower pace in our exploitation of energy. The phenomenal world is a complex process of energy transformations, and so is life, and so is the technological world.

The concept of energy is implicit in many ancient writings. The Vedas of India, which date back to more than three thousand years, speak of the divine as *cosmic energy*. Ancient Greek thinkers used the term *energeia* to mean activity. Medieval thinkers felt that there was some aspect of motion that remain unaltered when changes occurred. Soon after the rise of modern science in the 17th century, scientific thinkers tried to quantify this invariant feature of motions. Gottfried Leibnitz said that the quantity mv^2 which remained unaffected in motion. If a body stopped in an impact, its vis viva became part of the motion of the constituents of the body. Other thinkers had other views, and interesting debates ensued on the matter.

Later developments

Energies if the operation, efflux or activity of any being: as the light of the Sunne is the energie of the Sunne and as every phantasm of the soul is the energy of the soul.

– Henry More (1642)

Henry More's statement above has one of the earliest definitions of *energy* in English. For a long time the notion of energy was not clearly defined even the world of science. By the beginning of the 19th century, the *vis viva* of Gottfried Leibnitz was named *energy* by Thomas Young. This was the first quantitative definition of energy.

In the first decades of the 19th century Sadi Carnot made a thorough analysis of steam engines and their work-output, and studied the relationships between heat and work, paving the way for the science of thermodynamics. But it was largely from the work of William Thomson (Lord Kelvin) and William Rankine in the second half of the 19th century that the word became current in scientific literature in its modern sense,

Associated with *energy* are the ideas of work, force, power and action. One writer in the 19th century used the last three words in just two sentences in connotations that have nothing to do with their technical meanings:

"It is *energy* - the central element of which is will - that produces the miracles of enthusiasm in all ages. Everywhere it is the main-spring of what is called *force* of character, and the sustaining *power* of all great *action*.

A major difference between science and non-science

is this: .In science, terms are precisely defined. Hazy or ambiguous use of words is not permitted. On the other hand, in poetry and literature, extended and metaphorical uses of words are permitted, even encouraged because not precision but enjoyment is the goal. The poet is not there to inform us of something, but to provoke reflection and imagination. That is why one cannot do much with beautiful poems. But much can be accomplished in practical terms, and needless debates can be averted when well-defined terms are used.

Work and energy

The real essence of work is concentrated energy.

- WALTER BAGEHOT

In technical science the terms work and energy are related. In physics, work is not something you do for a living, nor a task to be submitted to someone. Work is what leads to energy. It involves a force that moves bodies. There is no work in unchanging motion. There is no work when the effects of forces cancel out. But when forces act and displacement occurs, then we say that work has been done.

Who does the work? Not this man or that woman, not this animal or that machine, as one would say in common parlance. Rather work is done by a forces on bodies. So we talk about the work done by weight or by friction, by the push of the hand or the tension of the string.

When a body falls, work is done on it by the earth's pull (gravity). When a speeding car screeches to a stop, work is done by the car against the frictional force. When the cable of the crane lifts a load up, work is done on the load by the tension in th cable. And so on.

Every time work is done on a body, the body acquires energy. Here then we have a technical definition of energy: an entity that arises in a body when work is done *on* it. Likewise, if the body were to exert force and cause displacement, i.e. if work is done *by* a body, the body will lose energy. Work is the mode for transfer of energy.

In other words, bodies gain or lose energy according as work is done on them (by an external force) or by them (by their own exertions). Thus, when a spring is compressed, work is done on it; so it gains energy. When a compressed spring bounces back, it does work, and so loses energy.

Since forces and displacements pervade the world, work is being done all the time, and energy transfers are ceaselessly happening. Never a workless moment for the tireless universe. This means gain in energy or loss thereof for the various bodies in the world. Creatures may work by day and rest by night, but the world keeps working for ever.

Units of work and energy

James Prescott Joule performed experiments which made possible the formulation of the full concept of energy.-

- **HENRY JOHN STEFFENS**

Like force and speed, work and energy are physical quantities. This means we need to define units in terms of which they can be measured. In terms of these we can compare the *amounts* of work or energy involved in processes. There are, several (equivalent) units for the measure of energy. The energy unit in the international scientific system is called the *joule*, abbreviated as J. Another commonly used unit of energy, especially of heat energy, is the calorie (Cal). In the technological world in which we live, energy (especially electrical energy) is measured in units of kilowatt hour (kWh). A kWh is simply 3.6 million joules.

We need energy for our survival as biological beings: our hearts and lungs, blood and brain are constantly doing work. And we move around too. And as technological creatures we also use (if not need) energy for heating, cooling, and lighting our homes, for transportation in vehicles, for entertainment (TV, radio) for cooking and refrigeration, for communication (telephone, fax) etc. All this can add up to quite a bit: about. As we shall see presently, all this energy is obtained from a variety of sources.

The rate at which energy is expended or consumed is relevant in practical contexts. This rate is called *power*. Power is a measure of how much energy is gobbled up or released in a unit of time (second). The international scientific unit of power is the *watt* (W). 1 W = 1 J/s. Thus when we use a 100 watt lamp, we are expending (electrical) energy at the rate of a hundred joules each second. A hummingbird spends about 0.7 joules a second as it flutters, while a supersonic jet guzzles almost 12,000 joules a second. No wonder we can't take a ride on the hummingbird to fly us quickly across the seas.

Kinetic energy

Soft is the strain when zephyr gently blows,
And the smooth stream in smoother numbers flows; But
when the loud surges lash the sounding shore,
....   **- ALEXANDER POPE**

Gentle breeze, steady streams, wild waves, every motion is a manifestation of energy. Energy appears in many forms, as we shall presently note, and one of its most common forms is motion.

Ordinarily we look upon motion only in terms of change in location. But when an ant creeps, a runner runs, an eagle flies, the pendulum swings, the wheel turns, the rocket zooms, and the planet moves: all display a burst of energy. We call this their *kinetic energy*. When Descartes spoke of matter in motion, he was recognizing the universality of kinetic energy. From microcosmic minuteness to grand galaxies, every bit of matter, be it a speck or stupendous, has kinetic energy with respect to some reference frame or another, for there is naught that is at absolute rest.

Kinetic energy has been precisely defined in terms of the mass of the moving body and its speed of motion. If a mini car and a huge van are both moving with the same speed, the huge van has far more kinetic energy than the tiny one. On the other hand, the faster any of the cars go, they greater will be its kinetic energy.

During the 19th century it was thought by quite a few thinkers that the energy of motion (kinetic energy) was the only form of energy there is: or at least the only form in terms of which all energy modes can be expressed. Heat and light and electricity could all be reduced to some kind of motion or another. This was the kinetic view of nature: a natural outgrowth of the Cartesian sweep of matter in motion.

Dissipation

To what purpose is this waste? **- NEW TESTAMENT**

We give a kick to the soccer ball; it flies and rolls and comes to a stop. We give a push to the child's swing; it's oscillations gradually weaken and bring the swing to a halt. We spin the wheel on its axle, and it too slacks in speed and loses its rotational motion. Even the furious meteorite that comes zooming into our atmosphere slows down, losing speed. Every moving body on earth slows down and stops, sooner or later. Put differently, all kinetic energy disappears.

What happened to the kinetic energy the ball received from my kick, the swing got from my push, and so on? Where did it all disappear? The answer is simple: the kinetic energy in these cases was dissipated by friction. Yes, it was lost as kinetic energy and became frictional heat. Only in the case of the meteor is this glowingly obvious, for the intruding stone burns bright and is spectacular as a falling star. The ball and the swing lose their energy to friction too, but the heat generated is so slight it is barely perceptible.

Frictional dissipation going on all around, for all motion here below is on surfaces: solid land, liquid water, or gaseous air. We can minimize, but we cannot eliminate friction altogether. A little grease perhaps at the point of contact from where the swing hangs, an ice floor for the hockey putt, will help reduce friction, but not do away with it.

Beyond the blue skies where no air pervades, the kinetic energy of a moving body is not dissipated. So the moon revolves and planets orbit without losing kinetic energy.

Near the earth, even at great heights where air is thin, there is enough friction to slow down satellites and spiral them down to a splash. Like a devouring pit, friction is gobbling up all kinetic energy.

Potential energy

Nature is often hidden, sometimes overcome, seldom extinguished. **- FRANCIS BACON**

Though kinetic energy is often dissipated into frictional heat, this is not always the case. When I kick a ball upwards and it lands on a terrace high above, the energy I imparted to it has ceased to be kinetic, but neither is it lost as heat. Rather it lies hidden within the body and can become kinetic again when it has a chance to roll off from the terrace. But in its silent static form the ball does not display the energy stored in it. We say it has *potential* energy.

Many bodies and systems in the universe have

potential energy: stored energy which may be released in different ways. Let us consider two simple examples first. When a pendulum bob is hanging with the string in the vertical position, it has no potential energy, nor kinetic. When it is displaced away from this position, it acquires some potential energy. Now if it is let it go, the potential energy is gradually transformed into kinetic energy until at the lowest level all the potential energy is transformed into kinetic energy. As it swings towards the other extreme position, the bob slows down, i.e. it begins to lose its kinetic energy and gain potential energy until it comes to a stop. Thus we see that as the bob oscillated back and forth, its energy is being continually transformed from the potential to the kinetic and vice versa. Note that the pendulum is most stable when the string is vertical, i.e. when the potential energy is least.

Consider a spring. In its normal state it has neither potential nor kinetic energy. When it is compressed or stretched, it is given some potential energy. When released, the potential energy is transformed into kinetic energy, as with the pendulum. Here again, the spring is stable when it is neither compressed nor stretched, i.e. when its potential energy is minimum.

Two facts emerge from these examples: In oscillations there is continual interchange between potential and kinetic energies; second, the system is in a stable condition when the potential energy has the least possible value.

There are many other situations where also energy is hidden. Thus, a matchstick has potential energy too, and this can be released as heat (fire) when it is struck. A candle stick, gasoline and a log of wood also have energy hidden in them, energy which can be released as heat and light.

If kinetic energy is like cash, directly negotiable, potential energy is like balance: safe and secure, retrievable under appropriate conditions.

Forms of Energy

To every Form of being is assigned....
An active Principle.   **- WILLIAM WORDSWORTH**

Like the soul of religions, energy has never been recognized in its unclad glory, but it manifests itself in countless different garbs. So there is heat and light and motion, sound and electricity too. All these are active principles: energy of one kind or another. From all these work can be done.

Energy manifests itself in two essentially different forms: First as motion which may be ordered or random; then as   waves, material or immaterial. These manifestations affect us in ways that serve our basic needs for survival, and also add richness to our lives. For example, ordered motion or mechanical energy serves us in locomotion; random molecular motion affects us as heat and warmth. Material waves include water waves and the vibrations of strings and air columns which produce sound and music. On the other hand, there is a whole range of immaterial waves which strike us as light and radiant heat.

The world of perceived reality is thus made up of energies in various forms. These different forms of energy create different impressions. We go through the pains and pleasures of life without pausing to consider the energy aspects of the world around us. But they are there at every turn, sometimes blatantly obvious, sometimes not so apparent, and in other contexts stealthily hidden, as it were, waiting to be released when opportunities arise.

Music can only be experienced in one of its countless manifestations. There is no such thing as music *per se*. It is always this song or that melody, this instrument or that chorus. So too there is no such thing as energy *per se*. It is always in one of its several manifest-modes. We observe and experience the various manifestations of energy like we enjoy different pieces of music.

Energy transformations

Look abroad thro' Nature's range
Nature's mighty law is change.   **- ROBERT BURNS**

But there are important differences with the music analogy. Music cannot be measured in units, as energy can. Nor can one piece of music be changed into another in the way in which the energy forms can. Heat can be converted to motion (kinetic energy), motion into electricity, electricity into light, light into sound, and so on. Indeed such transformations are the root of perceived reality. This whole experience we call the phenomenal world is merely an endless complex of energy transformations.

Indeed, energy transformations are endless, and they are complex too. Even the most trivial episodes in the arena of cosmic history involve complex energetic changes. It is beyond our comprehension to fully visualize the extraordinarily complicated ways in which energy changes occur in the universe, either partially or totally or along multiple channels, and give rise to phenomena. Consider snapping a finger with muscular effort. The energy for that came from biochemical reactions in the ingestion of foods, which came from molecules (proteins, fats, etc.) whose source is the green kingdom which captured some energy from sun light, which energy came from the core of the sun as a result of nuclear reactions which occurred, as per our calculations, some ten thousand years ago.

Yes, this is a most fascinating story: this conversion of energy from the sun's deep interior several thousands of years ago to its climax in the snapping of my fingers. And then it goes on, for that mechanical energy became sound and got dissipated in the air, giving a slight boost to a few molecules. What a mind-boggling chain of transformations! And this is an insignificant fraction of the countless transformations occurring at all scales of perceived reality.

At every pause we take to reflect on the roots of perceived reality we never cease to wonder at the magnificence of it all, in scale and in variety, in complexity and in simplicity. How fortunate that we can fathom some of this and wonder at it all!

Conservation

I know that any weed can tell
And any red leaf knows
That what is lost is found again
To blossom in a rose.   - LOUIS GINSBERG

Energy transformations are not random, nor in arbitrary measures. Like exchange rates in international currency, so much heat energy is equivalent to that much mechanical energy; that much mechanical energy is equivalent to so much light energy, and so on. So when a certain amount of energy is transformed into energy of some other form, the resulting amount will be clearly defined. In other words, like matter, energy cannot change forms, but cannot appear from nowhere, not disappear into nowhere. We say that there is a principle of energy conservation.

Then why all this scare about our running out of energy if the total amount of energy will forever remain the same in the universe? However, energy in one form may become less available than energy in another form. It is easy to switch on the light, converting electrical energy into light, but this light cannot as easily be transformed back into electricity. I can exchange a dollar into an ice cream cone, but once I have done this, it is not as easy to change it back into a dollar.

Aside from energy transformations occurring in the physical world, we bring about transformations for our specific purposes. Primarily, depending on the context, we utilize only four forms of energy: heat, light, locomotion, and sound. Technology consists essentially in devising ingenious and effective channels for bringing about appropriate energy transformations.

Matter and energy

$E = mc^2$ -ALBERT EINSTEIN

No formula of physics has gained greater glory and renown than the simple-looking one above with which the name of Einstein is associated. It has become the hallmark of 20th century physics. Let us reflect a little on what it says.

We have seen that energy is the insubstantial dynamic dimension of the physical world: that which is manifest as light and sound, as heat and motion or is simply held trapped in positional constraints and in molecular configurations, ready to be released when provoked. Matter on the other hand, is the static substantial feature of the world, concentrated and massive, localized in space, exerting nevertheless long range forces on other pieces of matter in the universe.

So we have a world of matter and energy, of substance and action, one causing us to feel we are in a solid tangible world, while the other making us feel the sensation of heat, joys of sound, and the effulgence of light. Matter undergoes umpteen transformations, while retaining its total mass; and energy does a very similar thing too, maintaining its quantitative integrity also.

These two dimensions of perceived reality, one so different from the other, are in fact two aspects of one and the same entity. We may regard matter as a concentrated form of energy characterized by its *momentum* (*mass and velocity*); and energy an insubstantial manifestation of matter. Matter may be converted into energy and energy into matter, violating conservation principles governing each. Einstein's neat little formula says that there is a precise quantitative equivalence between the two, meaning that for a given amount of matter there is a precisely equivalent amount of energy and vice versa. In other words, if a certain amount of matter is transformed into energy, erasing it from the material world, then a precisely determined amount of energy will emerge. Matter can be annihilated with the consequent production of energy. Equally amazing, subtle energy can materialize into gross matter: a gush of dazzling light can, in principle, become a little massive speck. There is only *momergon*. Which is manifest as matter or as energy.

For millennia people have known this world to be material. Philosophies had developed to the effect that there is naught but gross matter in the universe. Those who held this view were called materialists, with a touch of moral repugnance. Perhaps this was because matter stood for flesh and associated carnality, in opposition to spirit and its associations with the Divine.

In any event, old time materialists will have to concede they were wrong. The world is not pure matter. It has energy too.

Coming back to the Einstein formula, it has earned its legitimacy in countless ways, not the least of which was the Hiroshima horror and the lighting of many towns and cities by electrical energy generated by the annihilation of matter. More spectacular still, we have come to know that this decimation of matter with the consequent release of radiant energy has been going on for astronomical ages in the core of the innumerable stars that adorn the nocturnal sky, and right in the heart of our own blazing sun. Every bit of sunshine we bask in comes at the sacrifice of some minute mass in the body of the sun. Slowly but surely, the sun is depleting its massive abundance by turning it all into heat and light and other radiations that perennially escape away, abandoning their place of origin forever.

Zero-point energy

For something in the depths doth glow
Too strange, too restless, too untamed.

- MATTHEW ARNOLD

The atoms in a piece of solid are in a state of vibration because they have kinetic energy. If we give the body some heat energy, its atoms vibrate faster. If we cool it, we are taking away energy from the atoms, and they vibrate more slowly. Now suppose that we cool the solid more and more, going as low as we possibly can. The atoms become more and more sluggish, becoming some long distance marathon runner who is so exhausted he barely has any energy left. So, perhaps by reaching the lowest possible temperature there is, we can bring the atoms to absolute rest?

Not really. Walter Nernst showed that it is impossible to reach the theoretically lowest temperature. According to Nernst's theorem, at the so-called absolute zero temperature there is perfect orderly motion rather than perfect rest of the atoms. Later developments in quantum mechanics further confirmed this result.

Energy, the life-breath of the universe, will never by annulled. There will always be a residual heartbeat even if the universe were to reach asymptotic chillness. Inactivity may be for the lazy, rest for the tired, and inertness for the lifeless. But a modicum of energy will always be associated with the atoms and the ultimate oscillators of the universe. Here is a root of perceived reality that has been unearthed by scientific inquiry: We live in a world that has been keyed to innate dynamism.

In a little poem George Herbert said that God poured

218

on man all the blessings he had in a glass, withholding
rest, saying:

> Yet let him keep the rest
>
> But keep them with repining restlessness
>
> Let him be rich and weary, that at least,
>
> If goodness lead him not, yet weariness
>
> May toss him to my breast.

Maybe this is what God said about the universe at large

Vacuum fluctuations

And nothing brings me all things.

- SHAKESPEARE (Timon of Athens)

As mentioned earlier, the most powerful material
microscopes do not suffice to fathom the smallest
frontiers of perceived reality. For this we need all the
penetrating power of mathematics. Equipped with this,
theoretical physicists have explored the deepest roots of
the material world and dug even further into the hidden
recesses of nothingness. And they have found, in
complete contrast to what common sense might tell us,
that there is activity galore in emptiness. Yes, in the
intangible sea of vacuum there are fluctuations like you
won't believe it: the appearance and disappearance of all
sorts of energy-bundles. This is what emerges from what
is known as *quantum field theory*: the language and

219

framework of one dominant tribe in current fundamental physics.

No, this is not fantasy, not the magical concoction of mythology where the strangest episodes are permitted, where beings may come and go by the whiff of a thought. The sort of thing I am talking about has grown from the solid soil of serious physics. The ideas rest on esoteric concepts like the electron's intractable *self-energy* which can swell to incalculable proportions, and on sophisticated mathematics which involve terms like the *renormalization of infinities*. From it all emerges a picture that is fantastic and fascinating. But more remarkably and importantly, these notions account for a whole range of experimentally measured data to an impressive degree of decimal places.

Physicists have found that energy does not vanish even where there is absolute nothingness. Like writings on a blank blackboard, these *virtual* (ephemeral) *particles* appear and are promptly erased, as if strewn and sucked back by some cosmic pump. This is possible because at the underlying levels there are inherent latitudes as to the strictness with which matter and energy are conserved. For very short time intervals, as we shall see later, minor violations are permitted in microcosmic processes.

The universe from nothing

From nothing I was born, and soon again
I shall be nothing as at first. - **Greek Anthology**

The kind of dynamic vacuum mentioned above is no irrelevant quirky behavior in some remote corner of the universe. It is not a stray event like a comet or a supernova which are interesting but not regular features of the observed world. Rather, if current cosmological models have any merit, virtual particles in the nether world are the ultimate culprits which are responsible for the emergence of the universe. They are what gave rise to this material universe of ours, and in quite unexpected ways too. It all happened some ten to fifteen billion years ago, they say: when all of a sudden a pre-universe of nothingness where neither space nor time, neither matter nor physical law existed, burst forth all of a sudden as the famous Big Bang, trumpeting as it were the birth of a universe, and initiating the chain of events that have led to you and me, and to a zillion other interesting things with the slow passage of time.

What a burst it was! In an unimaginably brief interval of time that lasted for 10^{-30} seconds, a tiny, tiny bit of this violent vacuum exploded to 10^{50} times its size, its stupendous energy stored in what has come to be called the Higgs field. It is this Higgs field that blew the space into a sphere of enormous radius. When the expansion became a little steady, there was a *symmetry breaking* at which instant particles and antiparticles materialized in abundance. Thus was born our physical universe. This picture has all been calculated, formulated, and debated by physicists as a possible scenario for the birth of the baby universe in international conferences and in papers in prestigious journals. It has even permeated into popular books.

Now one may ask with a touch of skepticism, is this what really happened? Is this how the universe came to be, and not in seven short days by a fiat from the Creator? Who can tell? But this surely is the sketch of cosmic genesis, as suspected by current cosmologists. They go on to say that ours may be just one of many other universes thus formed, like bubbles in the vastness.

9

WAVES:

RHYTHMS IN THE WORLD

The world is full of poetry – the air
Is living with its spirit; and the waves
Dance to the music of its melodies.

G. C. Percival

What are waves?
When I play on my fiddle in Dooney
Folk dance like a wave of the sea.   – Wm Butler
Yeats

Standing firmly on a wet sandy beach, facing the vast blue ocean, one can feel the incessant surges of majestic waves incessantly that lashing out on our feet. They do this on the long shoreline, like a furious army charging at a defiant enemy, but only to die away into foamy puddles, and meekly recede back into the ocean.

They are noisy, those waves, splashing on and on with never-ending persistence. They have been doing this day and night, summer and winter, since time immemorial: long before humanity emerged on earth!

They don't care if people swim near where they kiss the land, they don't care if petty creatures make a living here and there in the waters. Nor do they care if their fury is recognized, heard or measured.

What strikes one most about waves is their periodicity, routine repetition to and fro. The waves we experience on the beach come from somewhere in the sea, and were transported across the body of water, starting somewhere, reaching somewhere. They are motions *in* the ocean, not *of* the ocean. It is important to note that a wave is the propagation of a periodic disturbance from one region of space to another. This disturbance is motion, and motion implies energy. So we may look upon waves as a mechanism for carrying energy from point to point. This is a function that waves serve: the transportation of energy from place to place, often faster and more efficiently than other modes.

Parameters associated with waves

Oceani fluctus me numerare jube: **You bid me to number the waves in the ocean.** — MARTIAL

If we wish to play the game of fruitful physics, we need to measure, we need quantitative descriptions. What this means is that we should attach numbers to waves. Let us call one full sigh of up and down a complete cycle of the wave. Now we can talk of how long it takes for a wave to make one full cycle. This is the *period* of the wave. A wave that takes two seconds for a rise and fall has a period of two seconds. We may count the number of cycles in a second. We call this the frequency of the wave and measure it in units called hertz (Hz). If a wave takes two seconds per cycle, it makes only half an oscillation each second. Its frequency is one half hertz.

The distance a wave travels during one full oscillation is called its wavelength. Thus, if a wave whose wavelength is 10 cm it means that it travels this far during one full oscillation. If this wave has a frequency of 60 Hz, it will travel 60 x 10 cm = 600 cm in a second which would be its velocity. More generally, the velocity of a wave = its wavelength times its frequency.

We measure how high the wave rises and call this its *amplitude*. If the wave rises to a maximum height of two meters, its amplitude is 2 m. We may reckon how far the disturbance travels in a second.

One speaks of a wave's period and frequency, amplitude and velocity. These provide useful quantitative descriptions of any wave we may be interested in. Numbers are not just for the tagging: they mean a good deal more.

There are many more numbers and formulas associated with waves. They need not concern us here. But there is one mathematical aspect of waves that deserves mention. It is what they call the *wave equation* in technical jargon. The wave equation is an expression in the compact symbolism of mathematics of how the measurable elements of a wave are interconnected. It is a powerful mantra in the esoteric language of mathematical physics. If one knows it and is skilled in manipulating the wave equation, one cannot useful information and interesting insights, and predict the course and constraints of a wave.

Waves: longitudinal and transverse
There are only two qualities in the world....

- SHAKESPEARE

Take a long slinky and place it on a smooth floor. Press it gently at one end and the compression will slowly travel to the other. A disturbance has been propagated from one region of space to another: we have a wave. Here the oscillations are *along the axis* of the spring, and so is the direction of propagation of the wave. Such a wave we call *longitudinal*, because its expression is entirely along a length.

Or again, consider a long row of pendulums on

strings of equal lengths, all hanging parallel from a long horizontal rod. Swing the first one gently, and when it hits the proximate one, this too will begin to swing, and on and on the disturbance travels along the line of oscillation. This too is a longitudinal wave. The disturbances are like it happens when a hundred matchboxes are kept standing side by side and the first one it tipped towards the next. One by one, like it says in the song, they all fall down: longitudinal propagation of a disturbance again.

Longitudinal waves arise in media where the elements are not in direct contact with one another, as in air. Sound consists of longitudinal compressional waves. Earthquake disturbances also involve longitudinal waves: rocks are forced to vibrate in the direction of propagation of the wave.

In waves on water, the motions of the medium are up and down while the wave itself is propagating horizontally. Thus, the lines of oscillation and propagation are *mutually perpendicular*. This is an example of *transverse* waves.

Consider transverse waves on a rope. We may make the rope oscillate up and down at one end, and the wave propagates along the rope. Transverse wave again. But we could also make it oscillate right and left. Indeed, the oscillation may be alone any line perpendicular to the rope. If the rope passes through a slit the line of oscillation is restricted. Then the transverse wave is said to be *polarized*.

Complex waves

The waves broke and spread their waters swiftly over the shore. One after another they massed themselves and fell; the spray tossed itself back with the energy of the fall.   **- VIRGINIA WOOLF**

We talked about the frequency and period of a wave. A wave with a single frequency and period is a pure wave, a Platonic ideal as it were. In the crass world of physical reality, things are seldom so simple: not on our scale of experience. Here we only observe overall effects, the jumbling up of myriad factors to make interesting blobs. That is what makes complexity the central theme when it comes to studying matters of immediate important to us.

Most waves in the world are complex: mixtures of many waves of different frequencies. They all add up as per an innate law of wave-addition and have one total effect. In the observed experience, it is not clear that the wave is a combination. It is somewhat analogous to a fruit punch: sweet and flavorful, a single homogeneous satisfying beverage, but constituted of a variety of juices of different tastes. So we see white light, for example, which is a combination of light of all rainbow hues. We hear a sound, a call or a sustained tune: we perceive it as one complete sound, or so. But in fact, it is made up of a great many sound waves of a great many frequencies.

The world of waves is always that way: never pure and simple, always complex and combined. What is equally remarkable is that we have means and methods for analyzing a complex wave into its component parts. This may be done at the physical level through all sorts of instruments: spectral analyzers they are sometimes called. They are somewhat like machines into which you shove some dollar bills and out come pennies and nickels, dimes and quarters, all neatly sorted out. Complex waves can also be analyzed through elaborate mathematical techniques known as Fourier analysis. Such analysis is essential in many contexts, as in the construction of filters for sound systems.

This is the sort of reductionism that some philosophers do not approve. Holists contend that reductionism is misdirected, narrow, and distorting. Some of them say that classical physicists, misguided in their addiction to the analytical mode vivisect Nature which is to be grasped in its totality. The point is, both are important modes. You need the holistic approach to uncover some aspects of perceived reality. You absolutely need the analytical, reductionist approach to know white light as made up primary colors or that gross matter is made up of quarks. Reductionism gives us profound insights about the roots of perceived reality. Holism is like tasting and enjoying food. Reveals to us knowledge of Carbohydrates and proteins.

Reflection

For the human heart is a mirror
Of things that are near and far;
Like the wave that reflects in its bosom
The flower and the distant star.   **- ALICE CARY**

Stand in a vast open field all alone and clap your hands. There is no one to hear, none to respond. Let it be region where a hill stands high, and you will hear the clapping of your hands again. What is this sound that bounces back? It is an echo, a prompt return of the sound to where it came from. Mysterious and magical it could be first, "a voice without a mouth, and words without a tongue," as Horace Smith once said.

All waves are reflected when they encounter an obstacle. A simple property that causes richness in our experience, It is the reflection of light from a polished mirror that enables us to see ourselves. If there were no reflections, we could never know how we look unless we have a photographs of ourselves. Echo is reflection of sound waves.

Sometimes reflection causes peculiar effects: like the mirage in a sandy desert creating images of trees below them at a distance and causing the illusion of a pool over there. Reflection traps light inside some crystals and then releases it with sparkle it as in a diamond. It is thanks to the reflection of radio waves from the upper sheets of the atmosphere that radio sets pick up broadcasts from distant lands. Our satellites reflect microwaves and make intercontinental TV and communication all so easy. Yes, there are myriad situations, emerging in nature and in human-made devices where the reflections of waves create scenes and spectacles that otherwise would never be there in the world.

Attenuation

The longest wave is quickly lost in the sea.

- R. W. EMERSON

You stand in a field and call a friend walking away at a distance. She does not turn her head. What happened to all the sound you made? It got dissipated slowly as it traveled the distance. The amplitude of a wave diminishes little by little: we say the wave *attenuates*, meaning its amplitude gets less and less along its course.

We feel the disadvantage of attenuation when we call someone in an open field. But if sound waves do not attenuate every secret whisper would be heard miles away, every conversation in a room would become public. There will be so much noise we wouldn't hear anything but suffer the pain of garbled loudness.

Attenuation makes water ripples die off in a pond. It is like the gradual slowing down of an oscillating pendulum which ultimately ceases oscillating when all its energy is lost to the medium.

There is no material medium between us and the stars, so their lights travel millions of miles, unattenuated. If there was matter in interstellar space that absorbed a little of the light passing through them, as some glasses do, then very quickly stellar radiations would die away, and there would be only pitch darkness in the night sky. Then we would never have known of the splendor and extent of the universe. We might have formulated a very different physics and cosmology.

Interference
When two Undulations, from different Origins, coincide either perfectly or very nearly in Direction, their joint

effect is a Combination of the Motions belonging to each.
- THOMAS YOUNG

Two cars from different directions happen to collide at a point of intersection. That is their end as moving entities. No more car moving beyond the point of the encounter. This is the nature of material entities: being arrested abruptly if they run into another.

But this is not so with waves. When a wave disturbance propagates through a medium it is disturbing every element of the medium along which it travels. Now suppose that another wave is passing through the same medium along a different path, and their paths happen to cross at some point in the medium. The two waves will then *interfere*, somewhat analogous to the unhappy collision of the cars. At the point of intersection, the elements of the medium will be agitated by the combined effect of the two waves. What is remarkable is that beyond the point of interference the waves will continue as if nothing had happened, i.e. as if they had not been struck their ways by another wave.

This property of waves plays an important role in the nature of the physical world. If light waves did not behave   this way, we will be unable to distinguish objects as separate entities, because the light from them would all be jumbled up. We cannot hear the sounds produced by single individuals because when the waves encounter other sound waves, they would be destroyed. All the light from the countless stars, like all sounds from different sources, would mingle and become one diffuse chaotic whole.

When two masses are combined, the result is a body whose mass is the sum of the combining masses. But when two waves of equal amplitude interfere, because of their changing effects on the elements of the medium, their combined effect could be either double the amplitude of both or simply nil, depending on the relative aspects of the waves when they meet. What this implies is that it is possible for two light waves to interfere, it is possible that at some points the result is total darkness; or when two sound waves interfere, at some points the effect could be total silence.

The phenomenon of light interference was put into observational evidence by Thomas Young by the first decade of the 19th century. He used pinholes and sunlight and the simplest of arrangements to unravel a most significant root of perceived reality. An arrangement by which he demonstrated the phenomenon of light interference (the double-slit experiment) has become a classic and is repeated to this day by countless students of physics in laboratories all the over the world. The quantitative side of Young's experiment enables us to determine the wavelength of light. This significant root of perceived reality was uncovered by modest experimental set-ups, costing very little. This was long before the era when experiments called for writing for grants and securing hundreds of thousands of dollars for equipment and assistants

Interference of waves causes many phenomena. The multicolored patterns in oil spills and colorful soap bubbles are due to the interference of light waves. By measuring the spacing of the colored patterns one can determine the thickness of oil spills. Films thinner than a billionth of a meter have been measured using light interference. At the other extreme, the diameters of stars have been measured by using interference also. By this method, the diameter of Betelgeuse in Orion has been estimated to be about 36 million kilometers.

Diffraction

Je plie et ne romps pas: I bend and do not break.

- LA FONTAINE

Throw a clay mug at a pole, it breaks at the obstacle. Hold a stick on the path of a ripple in a pond; the ripple bends around and continues. This ability of waves to bend around obstacles on their way is called *diffraction*.

The diffraction of sound waves is easily recognized. You may call someone on the other side of a wall. Sound waves diffract and reach the person. The diffraction of light waves is not as obvious. You can't see the person on the other side of the wall. The reason is that waves can only diffract around obstacles whose sizes are comparable to their wavelengths. Light waves have very short wavelengths. Walls are too large for them.

It is somewhat like our ability to jump over hurdles. Everyone can do it, but if the hurdle is too large for our height, we cannot show this human potential. After all, who can jump over Lake Ontario or the Alps?

Consider a small hole through which a narrow beam of light is made to pass. Were it not for diffraction, one will see only a bright spot on the screen. But what one observes is a series of bright and dark circles, resulting from the interference of the diffracted waves.

This is why when a point source of light is observed through an optical instrument (eye or telescope) the resulting image is never a bright point. Because of diffraction, we see a small circular spot around which there are dimmer bands. As a result, if two bright spots which are very close to each other are viewed, the images overlap into a single blur. That is why we cannot distinguish between double stars or planets and their satellites with the naked eyes. The degree to which an optical instrument can distinguish between two close point sources known as its resolving power.

Waves on strings

Take but degree away, untune that string,
And hark what discord follows!

- SHAKESPEARE (Troilus and Cressida)

Music is a major source of aesthetic delights. Many musical instruments are string-based. Waves set up in strings fixed at both ends are at the root of the music they generate. Musical sound arises from the superposition of one or more waves of discrete frequencies. This is possible with strings of finite lengths fixed at both ends.

Musical vibrations result from the interference of waves reflected back and forth from the fixed ends. The wavelengths depend on the lengths of the strings, and frequency on the tension with which the string is plucked as well as on the mass density of the string. This means that we may use strings of different thicknesses to produce different notes. Here we recognize a most interesting root of perceived reality: the beautiful music we hear played on violin or cello, piano or harp, results from the variety of waves (vibrations) that may be set up in fixed strings!

We all carry fixed strings in a little cartilaginous tuner called the larynx. This vocal box protrudes slightly as Adam's apple. It is also the entry to the lungs. Air from the lungs is extremely important, even as in any stringed instrument we need resonant air to vibrate sympathetically for sound to be produced. So we have the oral cavity which gets the air puffs from the vibration of the cords. By controlling the air puffs in different ways we can produce dull and unimpressive noise, intelligible talk, or the most sophisticated musical notes. Women's voices tend to be shriller because the mass of a woman's vocal cord is usually less than that of a man.

Electromagnetic waves

And God said, Let there be electromagnetic waves: and there were electromagnetic waves.

- BOOK OF GENESIS (*Scientific version*)

When it says in the Book of Genesis, "God said, Let there be light: and there was light," many profound truths of perceived reality are expressed: Light is universal, i.e. there is no spot in the world where there is no light. Light has been there since the very moment of cosmic birth. Without light there can be no life, here on earth or anywhere else. But for light we would be in permanent isolation from the rest of the universe, for it is light that establishes connections and communications among various entities in the universe. It is therefore insightful and appropriate to trace light to the very initiation of the world.

It turns out that light is only one variety of a more general kind of waves called electromagnetic (e.m.) waves. That is why practically every valid statement about light, and even truer if we replace the word *light* by *electromagnetic waves*.

The discovery of electromagnetic waves was a crowning achievement of (classical) physics. It is also an incredible story, wrought with the most astounding consequences to human life and civilization. In brief, this is what happened. By the middle of the last century, James Clerk Maxwell cast in mathematical form all the then known laws pertaining to electricity and magnetism: which were the fruits of a century of investigations by physicists. What this means is that Maxwell used plain paper and pencil (and his genius) to express the variety of electrical and magnetic phenomena that had been uncovered by careful experimental observations.

Two most surprising *theoretical* consequences followed from Maxwell's mechanical-mathematical model:

First, if electric charges oscillate (accelerate), then there electromagnetic disturbances will be produced which propagate as waves. Thus, the existence of electromagnetic waves was uncovered on paper using mathematics! Pause to think about this: electromagnetic waves which play a most powerful and ubiquitous role in 20th century civilization were discovered first, not in a laboratory, but through mathematical exploration of well-established physical laws.

The second result from Maxwell's analysis was that light itself is a kind of electromagnetic waves of which there are many. Today we are aware of a whole range of them, from penetrating gamma rays to long radio waves, through x-rays, microwaves, etc. All these constitute the *electromagnetic spectrum*.

It is one thing to become aware of a subtle root of perceived reality, and a different matter to recognize its physical existence. Not too many fellow scientists may even comprehend the new idea, let alone accept it overnight. It took more than two decades and the efforts of many before electromagnetic waves were detected in a laboratory by Heinrich Hertz. The detector used by Hertz in his laboratory in Karlsruhe for putting into evidence electromagnetic waves was very simple. It was nothing like today's complex experimental set-ups in huge establishments, requiring millions of dollars and involving hundreds of people. What Hertz observed were slight sparks: but what momentous sparks they were, with awesome potential for revolutionizing science and civilization.

X-rays

For any man with half an eye
What stands before him may espy;
But optics sharp it needs I ween,
To see what is not to be seen.   - JOHN THUMBULL

The discovery and study of each of the many subsections of the electromagnetic spectrum may be considered separately. Let us look into one of them: namely X-rays.

In the mid-1890s the *Daily Chronicle of London* wrote that "the noise of the war's alarm should not distract attention from the marvelous triumph of science, which is reported in Vienna." The war was the Boer War, and the triumph was the discovery of X rays. The war is now buried with the dead, but the exploration and use of X-rays have grown beyond anybody's wildest dreams.

That discovery was made on November 8, 1895 by Conrad Wilhelm Röntgen: a professor of physics in Würzburg, Bavaria. His researches on electrical discharges through gases revealed the existence of "rays" which could penetrate chunks of matter. With them, one could see what lies behind screens, beyond doors and beneath the skins that cover the body.

Röntgen was so puzzled by his discovery that he called them *X-Strahlen* (X-rays), meaning unknown rays. Using them, he promptly took a picture of his wife's hand, revealing the eerie bony structure underneath the soft, cushiony skin. She is said to have been frightened.

At the end of a public lecture, Röntgen showed to a curious audience the X-ray image of the anatomist Kolliker's hand. Everyone cheered tumultuously, and Kolliker said the rays ought to be called *Röntgen Rays*. It is known as *Röntgenstrahlen* in the German speaking world.

The news of the discovery spread far and wide through scientific journals, newspaper headlines and magazine articles. Not everyone understood its scientific import. Some thought it was a fad that would fade away, while others, with as much caution as naiveté, imagined its mischievous potential. One legislator, fine-tuned to Victorian morals, nervously feared that some naughty inventor would come up with a device (using X-rays) to enable peeping Toms in a theater to get glimpses of what lay beneath the garments of actresses. So he introduced a bill to ban X-ray opera glasses. Some clothiers advertised X ray-proof dress to protect people's modesty. Others hoped that vivisection would soon become a thing of the past in the study of human anatomy.

Within a year, almost a thousand scientific papers were published on X-rays. One writer who found all this to be a trifle excessive exclaimed in rhyme:

> The Roentgen Rays, the Roentgen Rays,
>
> What is this craze? The town's ablaze
>
> With the new phase of x-rays' ways....

X-rays were put to use for a variety of purposes. Their earliest medical application was in wars: The Italian army in Ethiopia (1896), the British army in the Nile Expedition (1896), and the U.S. army in the Spanish American War (1898) were the first to use X ray machines to examine the wounds of soldiers. Today everyone is familiar with their diagnostic value: to reveal broken bones, infected lungs, dental decay, in the detection of diseased tissues, etc. X rays were also used at one time to relieve arthritic pain, to cure fungus ailments, to get rid of pimples, to treat herpes, and to control tumors.

In less than a year after their discovery, X ray photographs were introduced as evidence in a court case in Denver, CO to prove that a young man who was suing his employer had indeed broken a leg when he fell off a ladder during work. Lawyers argued vociferously as to the admissibility of such "ghost pictures." Today, scanning electron microscopes using X rays are used in forensic science to find out if a suspect fired a gun.

X-rays have served in detecting art-fraud: With their aid one can find out the percentage of lead in a paint and find out if a painting is from an old master or a modern copy-cat whose paints carry much less lead. The Dutchman Hans van Meegeren was an expert forger the who had sold several "old Masters" to Nazi occupiers. He was exposed when his paintings were subjected to X ray analysis in 1946.

When Röntgen generated X rays in his lab, little did he know that the universe is bathed in X-rays emanating from myriad sources. In the 1960s astronomers detected a powerful X ray source in the constellation Cygnus: revealing a black hole in the region. In the 1990s intense X ray sources were detected near the center of our galaxy.

As to human capacity for X ray production, in 1995 two laboratories in the world (the ESRF in France, and the Argonne National Laboratory in the U.S.) generated enormously intense X ray beams - the most powerful till then. These enable scientists to probe directly into the heart of matter, exposing the structures of molecules and atoms, even as common photographs using light show images of things on our scale. In other words, X rays inform us about the minutest of matter units as well as the most distant cosmic concentrations. It is remarkable that the same tool can be used to probe the vary small and the very large.

X-rays are relevant in many contexts beyond looking for tooth decay and the apprehending airplane terrorists. Neither Roentgen nor anyone could have imagined that his discovery was the beginning of a new era in science: not just for its applications, but for the conceptual revolution it has wrought in our understanding of perceived reality: X rays have opened uo vast vistas of knowledge, revealed new facts relating to matter and energy, and engendered fresh insights into the nature of the universe. A scientific discovery is like the first sentence of a new novel that is written down. Not even the author can predict how the story will evolve.

Seismic waves

The earth shook, and the heavens dropped at the presence of God. - PSALMS (*lxviii*)

Earthquakes are unwelcome jolts to the stability of lands that cause damage to dwellings and harm to our lives? Many of them have occurred, in China and India, Alaska and Yugoslavia, Chile and Portugal and in other parts of the world as well. It is estimated that during the past four millennia more than ten million people have perished as a result of these intrusions into our daily lives.

When Norway was hit by a bad earthquake in 1657 (April 24) and a panic ensued, a wise and scientifically inclined theologian by the name of Millel Oedersön Escholt wrote a detailed analysis of the earth. When there was a terrible earthquake in Lisbon in 1755. Voltaire wrote a moving poem on the disaster questioning theism. Nicolas Desmarest wrote an essay on how earthquakes tremors are propagated. In the early 1880s a group of scientists began a study of earthquakes in which they used mathematical techniques for analyzing the data. An extensive theoretical analysis by Horace Lamb of the modes of oscillation in an elastic sphere, play an important role in boosting the science of seismology.

When Andalusia was shaken in 1889 scientific investigators rushed to the place to study the tremors and measured the velocity of shock waves in different soils. By investigating the travel times of seismic waves one probes into the bowels of the earth, locates the epicenters of earthquakes, and generates data which help us understand the motions of the earth's crusts.

When the string of a musical instrument is elevated from its normal position, and suddenly let go, oscillations result. So it is with the rocks in the crust, kept under tension for too long, and then released at some instant. Tremors follow. This is how seismic waves are generated. Armed with instruments geophysicists have been collecting considerable data on earthquakes. From these we have come to know that there are seismic waves of different kinds. Some are slow moving surface waves which, like waves on sea, move along the crust. Then there are the much faster moving ones, called primary (P) and secondary (S). Of these again, the P waves move faster, penetrating deep into the inside of the globe. These are compressional (longitudinal) waves. On the other hand the relatively slower S waves are transverse in nature, attenuating rapidly when they enter a liquid medium.

Seismographs measure even very slight seismic waves. From analyzing these data geologists determine with precision when and where an earth tremor originated. These instruments have been compared to X-ray machines because with their aid one can see what is going on inside a volcano.

Not many may think of Desmarest or Beno Gutenberg in the context of earthquakes. But they are among the investigators who first carefully studied and reflected upon the earth-shaking phenomenon.

Brain waves

Peut-être ferait-on bien des découvertes sur cette merveilleuse union de l'âme et du corps, si l'on osait en aller chercher les liens dans le cerveau d'un homme vivant. Perhaps one may make discoveries on this marvelous union of spirit and body, if one dared to look for connections between them in the brain of a living man.

- PIERRE MORAU DE MAUPERTUIS

Our world of experience is dependent on the functioning of the brain. That functioning involves complex electrical activities. These in turn generate subtle waves which were first noticed and studied by Hans Berger. These are electrical rhythms in the brain which, when recorded with instruments (called electroencephalograms or EEG) on a roll of paper, appear as complex wave forms.

When we analyze brain waves, we find that there are at least four kinds of them. First there are the *alpha waves* which are a background pattern common to all normal brains. These fast-moving waves with not too great amplitudes are very apparent when a person is fast asleep or just relaxing with eyes closed. These have been recognized as " sinusoidal resonance pulses in idle motor neurons." When one is under stress or agitated or intoxicated another type of brain waves, called *beta waves*, arises. These waves have still smaller amplitudes and travel much faster. Then there are the slowest waves, known as *delta*, which are clearly recognizable in the EEG when a person is in deep sleep. Finally we have the *theta* waves which come about when the brain is affected in some abnormal way, through direct physical damage or psychological shifts in personality.

A knowledge of these waves is useful in fathoming the mysteries of mind and thought. The patterns of brain waves in practitioners of meditation and in scientists have been studied. As a result of yogic exercises Swami Rama of Rishikesh produced all four brain waves simultaneously: a remarkable feat. Aside from recognizing meditative practices as more than exotic Eastern modes, scientific exploration of this kind exposes the physical basis of meditation techniques. This knowledge is also useful in the diagnosis of disease and wounds suffered by the brain.

Thus, waves are at the very core of our conscious

existence. There is so much rhythm in this world of ours, not just in music and in drumbeats, but in pulsating stars, in heart beats and in cerebral activities as well.

Gravitational waves

Nowhere in the Nature accessible to us do mass-oscillations of sufficient power occur to allow the resulting gravitational waves to be observed. -
Hermann Weyl

In 1916 Albert Einstein presented a paper to the Royal Prussian Academy of Sciences in which he showed that when the equations of general relativity are solved (even approximately), gravitation travels from one body to another in the form of waves. Two years later, in another paper, Einstein discussed the effect of gravitational waves on mechanical systems. At least one astronomer put forward the theory that gravitational waves interfered with light waves to produce effects like the red-shift of galaxies. In 1922 Arthur Eddington showed that gravitational waves must travel with the same velocity as light. These discussions instigated considerable theory-building and engaged many mathematically inclined physicists for at least three decades.

Gravitational waves could arise from two kinds of

sources: periodic and catastrophic. Periodic sources would include any body, say a rod, spinning around at a regular speed. Another example could be a spinning star. By astronomically catastrophic events we mean occurrences like the explosion of a supernova. Such events also generate gravitational waves. Gravitational waves are extremely subtle and cannot be easily detected. They are very poor energy carriers of energy.

Experimental work on the detection of gravitational waves did not begin until the 1960s. This involved sophisticated instruments. By the close of the decade it was known that detectors placed a thousand kilometers apart can reveal coincidences attributable to gravitational waves.

In 1974, astronomers discovered a pair of neutron stars orbiting each other. In principle such a system must be emitting gravity waves. As a result they should be losing energy, which means that they should be getting closer to each other as they orbit. This in turn should speed them up and decrease their period of revolution. Careful measurements carried out for more than fifteen years showed that this was indeed happening. This is taken as an indirect evidence of gravitational waves.

Efforts to detect gravitational waves are among the finest instances of the utterly disinterested spirit of the scientific quest. At this point it is very unlikely that the putting into evidence of the existence of these waves will be of any practical value. But that does not concern the probing scientists. Their only interest is to reveal and understand yet another root of perceived reality.

Matter waves

There throbs through all the worlds that are
This heart-beat hot and strong...   **- DON MARQUIS**

We look upon matter as blobs of mass, stationary or moving, localized at any instant at some point in space. Waves, on the other hand, are spread out entities, unconfined insubstantial oscillations. Thus matter and wave seem to be two distinctly different kinds entities.

But this is not altogether true. In the early 1920s Louis de Broglie proposed from theoretical considerations that with every material particle must be associated an intrinsic undular aspect: matter waves, as it were. They are not easily recognizable. They are unlike moving cars and tennis balls, or with anything on our scale of experience. But with the imperceptibly minute denizens of the microcosm - like electrons and protons - this wave aspect of matter becomes more than apparent in effect and significance.

Thus, as an electron travels from place it place, it is

not simply a speck of matter that zooms in the invisible world of fundamental particles like a planet or a football, but a tiny entity with an associated periodicity, crudely analogous to a creature with a heart-beat.

De Broglie's hypothesis was no empty speculation. He derived what the wavelength of such an associated matter wave would be, should such an entity exist. It would depend on the mass and speed of the particle in question in a very precise way, he stated. Matter waves are not a mathematical fiction: They have been observed through their effects. An intrinsic characteristic of waves is interference. If there are electron waves, they too must interfere. Not long after De Broglie's thesis was presented, the interference of electrons was experimentally observed. From careful measurements, the associated wavelengths were also determined. They turned out to be exactly what De Broglie had calculated.

So at the deepest levels of the material world, there is a vast sea of subtle surges, an invisible ocean, as it were, in which the particles would be like ships heaving on undular ups and downs. Because the matter specks and the associated waves are one and the same, there is no distinguishing between one electron and another: there are only overlapping clouds. Due to the wave aspect of matter, the microcosm is a whole different world.

10

SOUND: WAVES FOR ORAL COMMUNICATION:

And beauty born of murmuring sound.

– William Wordsworth

Sound: its variety

A thousand trills and quivering sounds
In airy circles o'er us fly...   **- JOSEPH ADDISON**

There is the serene chant of worship which uplifts the soul, and the magic of the mantra with its occult significance. There is the melody of music which fills us with delight and the hearty laughter which reflects a happy feeling. There is the cooing of the birds and the gurgling of the stream in rustic nature. There is the shriek of the frightened, the moan of the dejected, and the wail of the bereaved. There is the noise of the machine and the roar of the thunder. There is the secrecy of the whisper and the abrupt knock on the door. There is the chime of the bell and the bleating of the goat. There is the call of a familiar voice and the vigorous beat of the drum. One can go on and on, listing all the wondrous variety of

254

sounds that fill our world of perceived reality.

Every sound enriches life in a different way and connects us invisibly to our surroundings. Some evoke thoughts, some incite feelings. Some create emotions and others only jerky responses. Some give us delight, others cause pity. Some convey information, others nothing at all. Life itself is sound and fury, whether it signifies something or nothing.

Though seldom dramatic in its manifestations, sound is an extraordinary feature of perceived reality that adds depth and meaning in a hundred ways. Its rhythmic forms in poetry and prayer seem to put us in communion with some higher realm of reality, and its magnificent expressions as music has is the closest experience of the Divine we can get in a physical body.

Something this relevant to human life, this significant in the world of perceived reality cannot be ignored by the scientific quest. Its roots need to be explored.

What is sound?

Hark! now I hear them, - ding-dong, bell.
- SHAKESPEARE (*The Tempest*)

What is this ding-dong bell that we hear? What is this sound we experience, rejoice in and respond to? There are two aspects to this question: First, we may wish to inquire into what in the external world is responsible for what strikes us as sound. Second, what is happening in the internal world underneath of skull that produces this sensation. Let us not forget that perceived reality consists of two parts: reality and perception.

Physics is concerned primarily with the reality aspect, and physiology with the perception aspect. There is always a correspondence between the two which is what creates the world of perceived reality.

Sound is something immaterial in that there is nothing substantial at all that one can feel or see when one hears a sound. Sound is a consequence of vibrations brought about in the air around. These vibrations are longitudinal pressure waves, subtle enough not to cause any turbulence in the air. The vibrations could be set up by any vibrating body: a drum or string, bell or the vocal cord or anything that can vibrate.

Air is always disturbed to one degree or another by the motions occurring in it. By simply swinging a rope or a baton we can set up vibrations in air. But we do not hear every swing because our sensory mechanisms respond to only frequencies within a certain range. Elephants can hear low frequency vibrations. No human hand can shake a cane at twenty or more cycles a second, for that is the minimum frequency of vibrations to become audible. On the other hand, a mosquito can flap its wings this fast: which is why we hear the annoying hum of the beastie as it hovers in our immediate vicinity. Nor do we hear the much faster vibrations of atoms and molecules in a solid, for there is also an upper limit to audible frequencies: about 20,000 hertz. But mice can hear very high frequency waves.

What we perceive as sound are waves in a material medium: air, water, or solid, for something to vibrate.

Magic occurs when the waves reach our ears. The surface area of the ear-drum is barely a centimeter square, smaller than our thumbnail. When sound waves strike the eardrum, the latter begins to vibrate. Then through the wiring made up of neurons (nerve cells), powered by potassium and sodium ions, and assisted by fluids and bone-structure, the stimuli reach the brain where the physical processes get transformed into the experiential mode. We *hear* the sound.

Loudness and energy

I said it very loud and clear;
I went and shouted in his ear.
But he was very stiff and proud;
He said, 'You needn't shout so loud!

- LEWIS CARROLL

ROOTS OF PERCEIVED REALITY

The sweet whisper of a beloved and the firm order of the military commander are both sounds. The soft rustling of leaves and the roaring noise of a jet plane are sounds too. But there are differences in their loudness. Loudness is a feature of sound that strikes us most. Up to a point it is necessary for sound to be audible. Beyond that it may be necessary for clarity. Too much loudness is a downright nuisance for many people.

Recall that sound is a wave, and wave transports energy. From the perspective of physics the loudness of a sound is merely a reflection of the amount of energy a wave carries: the louder a sound, the greater the amount of energy it carries. Compared to the energy amounts involved in some other common contexts, the energy carried by sound waves is pitifully small. It is good that this is so, for the marvel of our ear can detect energy stimuli that are as low as a tenth of a quadrillionth of a joule per second. If, therefore, sound waves carried much larger amounts of energy, they would cause intolerable loudness, pain, and deafness too.

It is interesting to consider energy inputs of such infinitesimal magnitudes. There are processes that are minuscule in magnitude compared to what we are accustomed to. Even a meager candle burning for a few minutes puts out more (light) energy than does an entire orchestra playing for two hours. It has been calculated that if all the sound energy from the noise of a subway train were bottled up for a thousand years and converted into heat, we will barely have enough energy to warm up a cup of water.

Loudness is measured on what is called the *decibel* (dB) scale. The faintest audible loudness is taken as zero dB. The loudness of a city thoroughfare may be about 60 dB, of a speeding train about 80 dB, and of an aircraft propeller about 120 dB. The loudest sounds to which we are exposed carry a trillion times more power than the faintest. But such magnification in power does not create loudness-explosions by similar factors.

The energy carried by any wave depends on its amplitude. If the swings are greater, the sound generated becomes louder. With sound waves amplitudes are unimaginably small: of the order of the billionth of a centimeter. This says something about the sensitivity of our perceiving apparatus.

In energy terms, sound is modest in production and sensitivity. We expend only infinitesimal amounts of energy when we speak or shout. George Gamow once made an interesting calculation as to how much energy a professor spends when he delivers a one hour lecture. It turns out to be of the order of a tenth of a joule. This much electricity would cost less than a millionth part of a penny! The sensible moral we can draw from this is that it is not for the energy spent in producing the sound that we pay professors, but for the meaning and message in the sound produced. Sound carries message and meaning. Therein lies its magic.

Pitch and frequency

I heard a thousand blended notes,
While in a grove I sat reclined.

- WILLIAM WORDSWORTH

Musical sound of a single pitch is generally referred to as a note. When we hear a singer practice her scales, we recognize that it is an effort to produce as pure and clear a note as possible. It is not easy to produce sound of a single pitch. We have devices which can do this.

But what is this the pitch of a sound? Pitch refers to the frequency of a sound wave we hear. Thus pitch is the experiential dimension of how many times the wave oscillations occur in a given unit of time. In other words we seldom hear a pure note by itself. Most sound we normally hear is a complex of waves of different frequencies. All talk and noise, even the notes from musical instrument, are made up of several sound waves with different frequencies, perhaps not all in the same amounts. By the phenomenon of interference these different waves combine to form a single wave: we experience the combined effects of waves of various frequencies.

It was only during the 17th century that the correspondence between pitch and frequency was established. Many experiments were performed in this context. Taking the length of a string fixed at both ends as representing half a wavelength, the frequency of the sound was fixed. And the puzzling thing was, how could a string of a definite length produce different frequencies. It was Joseph Sauveur who first realized that a string could vibrate in various modes: he called these harmonics. Though his hearing was far from normal his work contributed much to the launching of the field of acoustics.

In the 18th century, string-vibration were theoretically analyzed by mathematical physicists, leading to many interesting results. At the same time the experimental study of sound vibrations on drums was also initiated.

In the 19th century a man named Karl Rudolph Koenig came to Paris after his education in Prussia. He worked for a violin maker and began to design musical instruments. Living in an apartment in the Île St. Louis in the heart of Paris he explored the nature of sound. Among the devices he constructed was a clock tuning fork with which he could determine the absolute frequency of sound.

The speed of sound

Yánai varum munné, mani ósai varum pinné: First comes the elephant, then the sound of the bell.

-TAMIL SAYING

C'est l'éclaire qui paraît, la foudre va partir:
It is the flash that appears, the thunder will follow.

- **VOLTAIRE**

The clap of thunder reaches our ears only a few seconds after the flash of lightning blinds our eyes: We conclude that sound takes time to travel even if light may not. When we talk to a person, even at the far end of a hall, unlike the thunder from the distant sky, our words seem to be heard right away. We do not see any lack of synchronization between the movements of the lips and the sound therefrom. This suggests that sound travels quite fast. The question is: How fast does sound travel? Another interesting question is: How does one come to know the speed of sound?

Once again, we go back to the 17th century when answers to such questions were first found. In the first half of that century, Marin Mersenne and Pierre Gassendi made experimental determinations of the speed of sound. In 1708 William Derham fired a cannon at one place and recorded how long it took for the sound to reach a point twelve and a half miles away. His result yielded a value of 1142 feet per second. In 1738 the French Academy of Sciences arranged to have cannons fired from Montmartre in Paris at half hour intervals, recording the sounds at Montlhéry eighteen miles away. Their data gave a speed of 336 meters per second: not very different from the currently accepted value.

What would happen if sound traveled at a much slower pace, say a few millimeters a second. Then when the professor is lecturing, students in the back rows will be hearing her much later: a few minutes later than those in the front rows! The thud of thunder would be heard perhaps an hour after the flash of lightning.

Both theory and experiments show that the speed of sound depends on the temperature and pressure the air, and more so on when the medium is solid or liquid.

In 1827 a bell was immersed at a point in Lake Geneva in Switzerland. When it was stuck some gunpowder was flashed above ground. More than thirteen kilometers away under the lake a huge ear-trumpet with a membrane was immersed from which a tube emerged above water. From the observed time difference between the instants the flash was observed and when the bell was heard at this distant point, J. B. Colladon and J. K. F. Sturm estimated the speed of sound in water to be about 1235 meters per second. J. B. Biot's experiments gave the speed of sound in iron pipes.

A remarkable experimental method in this context is the use of air columns for determining sound velocity. This experiment is still performed by students in laboratories. What is impressive here is that within the confines of a small room, indeed on a tabletop, using just a tuning fork and a hollow tube of adjustable length one can determine the speed of sound. Firing cannon balls is not needed any more, nor observations at points miles apart by different individuals.

These are not the sorts of things we take note of in our histories. They are not wars and victories, not political events or social upheavals; but they surely are adventures of the human spirit, conquests of the mind that enrich our understanding of the world.

Wavelengths of sound

She will start from her slumber
When gusts shake the door;
She will hear the winds howling,

She will hear the waves roar.

- MATTHEW ARNOLD

All this is possible only because the wavelengths of sound are not way too small. When we hear a baby crying in her room, sound has turned around the walls and reached us. As noted earlier, the sound waves diffracted around obstacles. One may conclude from this that the wave- lengths of sound waves are of the order of magnitude as the doors and the walls. Indeed they are.

When we strike the middle C on the piano, we create a sound whose frequency is 262 Hz. Recalling the relationship between wavelength and frequency (speed of a wave = wavelength x frequency) we may calculate that this corresponds to a wavelength of about 4.3 feet. The same note five octaves higher has a frequency of 8384 Hz, hence a wavelength of only 1.7 inches.

It is good that sound waves have wavelengths of a few inches and feet. Imagine that the wavelength of sound was of the order of a few millionth part of a millimeter. Then, we would not be able to hear a person calling us from right behind us, or the dog barking in the kennel or the prowler's stealthy steps, because the waves will not bend around to reach our eardrums.

This is a good example of how certain quantitative features of physical phenomena make the world make us

265

perceive the world the way we perceive it. It is not simply the physical laws that create the impressions we receive, but equally the numerical values of their measurable aspects.

Based on this some say that the world was created the way it is for us. Sound waves were sized for us to be able to hear the cry of the baby from the next room. Aside the fact that microbes survive without this benefit, this argument is as valid as the view that our noses were made with bridges so that we might be able to rest our eye-glasses on them.

Resonance

Some chord in unison with what we hear
Is touched within us, and the heart replies.

- WILLIAM COWPER

When we hear certain kinds of music we begin to nod or tap our feet almost instinctively. There are common meeting points, as it were, which prompt us to respond with sympathy. There is a physical phenomenon very similar to this, and we call it *resonance*.

Every system that can oscillate has a *natural frequency* (period) of oscillation. The pendulum is the simplest example. When made to swing, a pendulum of a given length oscillates with its natural period. So does a child's swing. The swing can be kept in motion only by being pushed now and again, or else it will come to a stop. Now, if the swing is given a periodic push at periods precisely equal to its natural period, the swing begins to oscillate with great energy: it is made to *resonate*. Resonance occurs when an oscillating system experiences a periodic force whose period is equal to the natural period of the system.

The effect of an applied periodic force may either oppose or add to the natural mode of oscillation of the system. This will depend on how well the applied force's variation changes rhythmically with the natural modes of the oscillating system. If the applied force changes in a manner quite different from the natural frequency of the system, the most it can do is to force the system to oscillate like itself. If. however, the applied force varies in a manner very similar to the natural mode of the system, then its effect will be to reinforce the oscillations. When this happens there is said to be resonance.

Resonance effects are common in physical structures. Attention must be paid to aerodynamic stability in bridge construction. In designing tall structures architects consider possible resonant effects due to winds. In 1952 light fixtures vibrated violently in a new office in Los Angeles because their suspensions were set into resonant oscillations by earth tremors. The transmission and reception of electrical oscillations through resonance are basic to the functioning of radio and television. When we tune a radio the circuit in our set resonates with incoming oscillations. The periodic steps of a rhythmic march of soldiers could cause forced oscillations of a bridge; hence a standard order for soldiers is to break step before crossing a bridge.

When a string fixed at both ends is plucked, it vibrates with its natural frequency. Waves may be set up in a cavity with air. The cavity has its natural frequencies: waves of well-defined frequencies can be generated in it with greater ease. Such a cavity becomes a resonator. Resonators are attached to most musical instruments. The vibrations set up by strings on the violin are picked up and reinforced in the resonating cavity, creating the melodious sound we hear.

Echo

Sweet Echo, sweetest Nymph, that livs't unseen
Within thy airy shell....- **JOHN MILTON**

As noted earlier in the open where there is a distant obstacle sound is reflected. We hear the report of a cannon fire more than once when there are hills around. Echo is sound that is reflected from a surface so far that the reflected sound arrives after the original one has died away. Whenever we generate sound the waves are reflected: from the ground, from the walls of the room, from the ceiling, and so on. But these reflected waves reach our ears far too soon to be heard separately.

When they bounce back from afar, they strike us as another wave, an eerie repetition of what we had sent out. It is a voice in the wilderness, like sound from a disembodied source. For physicists, this simply another interesting phenomenon, consequence of the reflection of waves.

We can live a peaceful life without echoes. But there are creatures for which echoes are essential for their survival. Lazzaro Spallanzani was a prolific scientific investigator of the 18th century. In 1794, when the man was a sexagenarian, he did some experiments with bats. He blinded the poor creatures and found, much to his surprise, that this did not bother them in the least in their flights! He jumped to the conclusion that the creatures had a sixth sense which enabled them to detect obstacles on their way. Others found that the bats did not fare all that well when their ears were sealed. The suspicion was around that the ability of bats to locate obstacles had something to do with their hearing. It took almost a century and a half before the matter was fully exposed.

269

Donald R. Griffin and Robert Galambos discovered that bats emit high frequency pulses whose echoes make them aware of obstacles on the way.

Porpoises can spot obstacles even in muddied waters thanks to the echoes of high frequency sounds they emit. In 1920s Paul Lengevin tried to utilize echoes from ultrasonic waves to locate submarines. Today the application of this idea has evolved: this is the basis of sonar and ultrasonic photography which enable us to chart the depths of seas and explore the sex of unborn fetuses also.

Applied to electromagnetic waves, this becomes radar which serves a wide range of purposes: from spotting speeding cars to detecting clouds and hurricanes. The simple phenomenon of echoes: once we know the cause of this perceived reality, a good deal flows from it!

Music and noise
Of all noises I think music the least disagreeable.

- **SAMUEL JOHNSON**

Johnson was of course not using the terms in their technical sense, for music cannot be considered a noise any more than that a random complex patch of lines may be regarded as a geometrical form.

At the experiential level the difference between music and noise is clear: we enjoy one and find the other not so pleasant. Both musical sound and noise are composites of waves, with this difference: musical sounds consist of discrete frequencies, while noise is made up of a continuous set of waves. It is somewhat like the difference between a can of pebbles and a can of flour. In the one case we can separate out the components, literally count them as so many; in the other case, it is one continuous pack of practically touching and indistinguishable parts. In a musical sound, such and such frequencies are present. In noise, practically all frequencies are present.

A spectrogram displays the component frequencies in a wave. The spectrogram of a musical sound would reveal straight lines corresponding to the frequencies present, whereas in with noise, we will find a continuous patch.

Consider a tone, say the middle A whose frequency is 440 Hz. This has higher harmonics as well. When this note is sounded in any musical instrument, the higher harmonics are also present to some degree. The relative amounts of various higher harmonics present when the note is sounded will depend on the instrument in question, because it is a function of the shape, size, material, mode, etc. of the resonating cavity. As a result, the same note sounds different when played on a violin or a clarinet, a piano or a flute.

This is equally true of the voice box or any source of non-musical sound. That is why there are distinctive differences between the voices of different peoples. Everyone has his or her own characteristic voice spectrum. Like the fingerprint, which is unique to the individual, each person has one's own voice prints: a vocal signature of our individuality. Its uniqueness has been used in forensics.

Music is soothing and pleasing; noise is not. At the tactile level, the situation is different: it is the smooth continuity of surface that causes the pleasing sensation of a soft touch. Discontinuities as on a surface with peaks and sharp protrusions, as in a bed of nails, cause unpleasant to touch.

Auditory system: the ear

Within a bony labyrinthean cave,
Reached by the pulse of the aerial wave,
This sybil, sweet, and Mystic Sense is found,
Muse, presiding o'er all the Powers of Sound.

                                        - ABRAHAM COLES

Ears add grace and balance to the face, for a human

face without ears would look odd and incomplete. The fan-like external protrusions, called *auricles* by anatomists, do more than serve an aesthetic function: They collect sound waves and ease them into the interior through a narrow dark passage, called the *external auditory canal*. This tiny tunnel, equipped with hair and wax to trap dust and other unwanted intrusions, leads to the eardrum (*tympanic membrane*) which covers a cavity (the inner ear) where three auditory *ossicles* (tiny bones), bearing the names of *malleus* (hammer), *incurs* (anvil), and *stapes* (stirrup) which lead sound waves into the inner ear. For the eardrum not to burst from the external air pressure exerted on it, we need air from the inside to balance this pressure. This is provided through the nose. Yes, part of the air we take in through the nose goes into the ear through the auditory tube and is meant to keep the eardrum from cracking.

The ossicles conduct the vibrations to the snail-shaped *cochlea* which contains a fluid. The fluid is agitated; this stimulates thousands of nerve fibers - about 24,000 of them. The neurons transmit the impulses to the brain. The rest, as they say, is mystery: for we do not (yet) know how electrical impulses get transformed into music and endearing calls.

In brief, mechanical vibrations set up in air first reach

the eardrum. These are transduced into electrical signals in the cochlea. When these signals reach the brain, the experience of sound is brought about.

To the inner ear are attached three little tubes, one horizontal and two vertical, containing a fluid. These are called *semi-circular canals*. Hair cells from these are connected to a nerve (*vestibular nerve*) which is wired to the brain. When movements of the body affect even slightly the fluids in the canals, the brain is alerted, and instructions are sent to various muscles in the body - from neck to toe - to do the needful to keep the body in equilibrium. Too much alcohol can interfere with the normal functioning of the semicircular canals: then stable walking becomes a challenge.

The ear is more than a hearing device. It also plays a role through an internal appendage in feeding data to the brain for not losing our balance. It has also been suggested that the semicircular canals may be responsible for our confinement - psychologically speaking - to a three dimensional space.

Voice box

Vox nihil allud quam ictus aer:
The voice is nothing other than beaten air. **- SENECA**

We talk and sing, whisper and shout. As stated

earlier, what makes it all possible is a little instrument in our throat, whose presence is indicated by the so-called Adam's apple. This little gadget is the marvel that enables us to communicate orally. It is the powerhouse whence emanate caring words of kindness and compassion, inspiring words of wisdom and ennobling sermons, and music, melodious and magnificent, that enrich our lives.

It is right there at the entrance to the respiratory tract, and in it are the muscle tissues we call *vocal cords*, attached on three sides. The crevice between the vocal cords is called the *glottis*. Vocal cords are subject to various tensions, thanks to a great many small muscles that control them. When air is forced out of the lungs, they are made to vibrate by the command of the brain. These vibrations create air puffs that resonate in the cavity of the mouth and the nose, generating all the wonderful sounds of the human voice ring in this world. If one is not impressed by this wonder and complexity, one will be immune to anything of significance.

There is more to the voice than production of sustained notes. It enables intricate speech with vowel and consonant sounds, the nasal and the guttural and the labial, their mixtures in words and continuity as sentences, long narrations and repetitions. The tongue, nose and lips, all play different roles here. We recognize these organs primarily as tasting, smelling and osculatory devices, but they are essential for speech too. We learn to manipulate them almost unconsciously when we learn to speak.

This sort of knowledge can never be obtained without subdividing and analyzing. No contemplation on the nature of ultimate reality can reveal these intricate wonders of perceived reality. Holism surely helps us to reflect, but reductionism is informational.

Our sensory organs receive physical inputs and the brain transforms these into a whole world of experience. The organs within the body serve the body's vital function, but there is one unique organ that proclaims our connection with the external world: it is the voice box. Unlike sensory organs which receive and vital organs which work silently and involuntarily, the voice box is a source and it is activated at our command.

The study of the mechanism of the human voice has been investigated for a long time. One of the first treatise on voice and hearing along with its history was published by a Giulio Casseri. In this work, Casseri compared the human voice system with those of other animals.

Ultrasonics

Fish say, they have their stream and pond;
But is there anything beyond?   - RUPERT
BROOKE

We are grateful for our faculties of sight, hearing, touch, smell, and taste. But these make us aware of only small fractions of the physical world. Often, our faculties are even deceptive in how they portray the world. But for ingenious instruments and imagination which is an extrasensory faculty, we cannot go too far in recognizing whatever is beyond our direct perceptual modes.

There are certainly many things not directly accessible to our sensory perceptions. There are, for example, pressure waves of frequencies much greater than 20,000 Hz and much lower than 2000 Hz which human ears do not detect. We call the higher frequency waves ultrasonic waves: sound waves which are beyond audible frequencies.

The trick is to cause something to vibrate very fast. Disks made of quartz when electrically charged do this, generating ultrasound. The device is called ultrasonic transducer. Ultrasound may be reconverted into electricity, activating computers which can then interpret the data.

In the 1940s sophisticated sonar came to be used for

detecting submarines. Since the 1980s ultrasound has been used to study the development of fetuses. Ultrasonic waves have found application in some unexpected contexts: like cleaning teeth and watches, mixing chemicals, even in welding, not to mention in the annihilation of kidney stones and brain tumors.

If light waves can be used for optical microscopes, why not construct an acoustic microscope using high frequency sound waves? This is what they did at the University of California in Irvine, with a billion Hz ultrasonic waves. When these are directed at tissues within the body, they bounce back as echoes. From an analysis of these echoes (with computers) one can gather information about the objects studied: which could be anything from cells to microbes and viruses, and some day perhaps the very genes that print out life patterns.

It was reported in 1994 that by injecting ultrasound into water with toxic chlorinated compounds, the latter were broken down. Thus one can use ultra-sonic waves to clean up certain pollutants from waters.

If someone of an earlier century had been told that the sex of an unborn child could be told without so much as touching the expectant mother's body, or that the precise location of an underwater ship could be determined without the use of light, it would have sounded like pure fantasy. If we had added that such things could be done by a form of inaudible sound, the statement would have been taken even less seriously. Such are the fruits of human knowledge.

Music of the spheres

Her voice, the music of the spheres,
So loud, it deafens mortals' ears;
As wise philosophers have thought,
And that's the cause we hear it not.

- SAMUEL BUTLER

The idea is very ancient, dating back in the Western tradition from the esoteric school of Pythagoras: that associated with the celestial spheres moving in harmony in high heavens is a cosmic music that fills the universe. Even as strings plucked in proper proportions produce the octaves on the scale, planetary motions in conformity with celestial arithmetic must create a divine music so went the reasoning.

The imagery is beautiful: the spheres were heavenly wheels on each of which stood an alluring nymph) that created a musical note. It is the combination of these notes that merged to form the music of the spheres. These became the Muses in Plato's school, and Christian mythology transformed these into angels and other celestial beings that form a celestial orchestra. It was believed that Pythagoras was one of the few who could hear the music of the spheres. Some thought that others, perhaps in the mother's womb or in early infancy recognized it, and that age and corruption soon deafened ordinary mortals to this universal harmony.

Aristotle thought heavenly bodies to be of perfect crystalline material. He feared the perfection of cosmic crystals would be shattered by loud notes in heaven.

Ancient Hindus believed the universe to be pervaded by mantras: divine poetry with spiritual power. These nuggets of eternal wisdom were revealed to the Himalayan rishis as Vedas: scriptural treasures of the Hindu religion. When Moses heard the Commandments and The Prophet Mohammed the Holy Qur'an, they too were privy to the cosmic music.

Underneath the poetry of the ideas there is the insight that as music is wedded to mathematics, so is our understanding of the universe at large. As scriptural wisdom is the key to understanding the nature of the universe (in the traditional mode), so was astronomy. The notion of music of the spheres was taken quite literally by Johannes Kepler in the 17th century

From current perspectives, it is difficult to imagine music in stellar space if only because we need an elastic material medium for sound. No one can sing on the moon, and even if one beats a drum with full force no sound will be created there for the moon is without any atmosphere. So all talk of music in the heavens, picturesque and poetic as it is, is not physics such as we understand it.

Yet, in a strange sort of way, the idea has a modern analogy. In 1966 astronomers discovered a microwave radiation that is cosmic in scope and that has been there ever since the birth of the universe. This is not sound, nor is music in the usual sense, but there is an all pervading vibration in the heavens, the Aum vibration in the Hindu vision: a notion that would have seemed strange and unacceptable to physicists of the 18th and 19th centuries.

11

LIGHT:

REVEALER OF FORMS & COLORS

Hail, holy light! Offspring of Heav'n first born!

- John Milton

Light from the sun

I saw myself the lambent easy light
Gild the brown horror, and dispel the night.

- JOHN DRYDEN

Have you ever been up   in the pre-dawn in pitch darkness in the woods or on a hill, on the bank of a river or on a beach, when there is no moon is in the sky: and wondered about the world around? Gradually, you begin to see the first rays of the sun, giving form and substance to things around. Soon the whole world is lit with sunlight, and everything becomes familiar and tangible, for in the darkness before dawn you felt all alone, even if you knew there was a landscape with trees, paths or whatever.

This has been happening for eons, this diurnal emergence of the sun all over the world. And for ages countless humans have woken up to the brightness of sunlight to begin another day of life and activity. The sun has been our first and perennial source of light. We cannot picture life on earth with no sun in the sky, a world of pitch darkness for ever and ever. It is appropriate that ancient cultures have been reverential to our central star. Akhnaten declared the sun to be the only God.

Today we flip a switch without even saying *Fiat Lux* (Let there be light!), and there is light in the room; or a flashlight can do this for us. Once there used to be hurricane lamps, candles and gas light to play a poor substitute for the sun. All we need is a little light to see what is around so we may not trip, so we may see the faces of people with whom we converse. Amazingly, even a small flame can do this for us.

What is light?
Oh say! What is that thing called Light,
Which I can n'er enjoy? - COLLEY CIBBER (*The Blind Boy*)

The question is relevant only to those who have experienced light. If we are born with normal eyesight, we may ask, What is the light that illumines the sky and all the world, this light that is a *sine qua non* for life and activity? Besides the tangible matter we see and touch, as mentioned in a previous chapter, there are also electromagnetic waves in this universe of ours. These are rapidly varying intensities of intrinsically locked-in electrical and magnetic fields that propagate through empty space with incredible speed. They appear in a wide range of wavelengths and frequencies. When electromagnetic waves whose wavelengths lie within a narrow band impinge on the normal human eye, they are transmitted to the brain which changes them into the experience of light.

Some ancients believed that light came from the eyes. As proof they said that we see nothing when we close our eyes. In a strange way, their view of light was not without merit. For, in truth, there would no light in the universe if there were no human (or other) brains to interpret them as such. If we may adapt from the Book of Genesis, we may well say,

> God said, Let there be Humans
>
> And there was light!

Eyeless human beings, endowed with an otherwise powerful brain, may still do a lot of physics, but they will not speak of light as something special. They might have become aware of the electromagnetic spectrum but would have categorized them differently. In this sense our science is anthropic, for the notion of light has no objective validity.

There electromagnetic waves whose wavelengths are less than a millionth of a meter. In terms of the nanometer (1 nm: a billionth part of a meter) these wavelengths range from about 420 nm to 700 nm: small, very small indeed; it is a wonder we have been able to measure them so accurately.

The human eye and its role

Golden hours of vision come to us in this present life, when we are at our best, and our faculties work together in harmony.   **- CHARLES F. DOLE**

Have you ever looked very carefully into someone's eye to discern its various parts, like the ophthalmologist does? Try this sometime, and you will notice in the eyeball the central *pupil* surrounded by the lighter *iris* around which is the *conjunctiva* on which you can recognize faint lines that are nerves. This is all that you can see directly. The conjunctiva is covering the outer layer of the eyeball known as the *sclera* whose front side is the *cornea* which holds a watery liquid called *aqueous humor*. Smack in the center of the eyeball is a transparent semi-solid called *vitreous humor*. Just beyond this is the choroid replete with blood vessels.

About eighty percent of the inner layer of the eyeball, mostly in the posterior region, consists of the retina which is endowed with cells that are sensitive to light of different frequencies. It has light sensitive pigments in its *rods* and *cones*. This is the messenger screen for it conveys the impinging light impulses to the brain where the magic occurs: electromagnetic vibrations become light visible with all its chromatic splendor. But only a hundred thousandth part of all the light energy reaching the retina gets interpreted by the brain. It takes about half a second for the brain to process the information. Neuron impulses take time to reach the brain.

The system is impressively simple in description, but wondrously complex in detail, and simply miraculous in what it accomplishes. It is this arrangement that turns vibrations into visual reality. All the beauty of forms and charm of colors, all the shimmer and shade and brightness we perceive is because of what the retina and the brain do. It is difficult to imagine that the cosmos in all its complexity would be plunged into demonic darkness if no eye responded to that narrow band in the electromagnetic spectrum. No glorious sunset or twinkling stars, no glitter or sparkle there would be, if the real vibrations palpitated unnoticed, unrecognized as light and color. How was the universe during those eons before life emerged? It certainly was not in this grand glory. Even when roses bloomed and leaves changed to their autumnal gold; it must all have been bleak and insipid to eyes that had no rods and cones. The multicolored fish we see in aquariums, the majestic rainbow, yellow sunflower and purple violets: all these were there for hundreds of thousands of years before the human retina evolved, but never recognized as such.

Let us not underestimate our role in the universe. Our little retinas add substantially to it all. They seem to make all of creation worth the time and effort.

White light and its components

Color est e pluribus unus:
From many colors is one.   **VERGIL**

For generations it was imagined that white light was the purest of all. But this is not so. White light arises from the mingling of every colored light there is. In the 17th century Isaac Newton uncovered this surprising root of perceived reality. He arrived at this discovery from observations with prisms. One can never get to the heart of physical reality without interacting closely with the phenomenal world. In his *Opticks*, Newton could firmly state that "Lights which differ in Colour, differ also in Degrees of Refrangibility," and that "The Light of the Sun consists of Rays differently refrangible." Now think of it for a moment: white light, the most colorless you can imagine, turns out to embody every color from violet to red that span the celestial bow!

In the beginning of the 14th century Theodoric of Vriberg published a tract in which he tried to explain the rainbow in terms of refraction and reflection, as did Archbishop Marco Antonio de Dominis three centuries later, as also René Descartes and Edmund Halley. Newton gave a mathematically more precise theory of the rainbow: others had been on the track before him But it was two more centuries before a complete theory of the rainbow would emerge. This was accomplished by George Airy in 1838.

Here is a scientific conquest which has broken the hearts of poets. There is a thrill in experiencing mystery which is crudely erased when logic and knowledge come to the fore. John Keats expressed it this way:

> There was an awful rainbow once in heaven:
> We know her woof, her texture; she is given
> In the dull catalogue of common things,
> Philosophy will clip an Angel's wings.

In 1853 Grassmann published a theory on the mixing of colors in which he explained that the human eye perceives three different aspects of light: *brightness, hue,* and *saturation*. The term color is a description of all these three components in an object (or image). It took more than a century before the three principal pigments in the cones of the eyes were identified. Hue refers to the basic colors of red, blue, and green, while saturation is a measure of how pure or uncorrupted a hue is.

Light absorption and colors

All sorts of colors, the which on earth do spring
In goodly colours gloriously array'd.

- EDMUND SPENSER

The world is filled with chromatic splendor. Green grass and multi-colored flowers; brown mud and blue skies, yellow birds and red sun at sunset: one can go on with the variety of colors in various shades, seven of which are displayed majestically when the colossal rainbow arches in the sky. To a dog, this same world would be as drab as actions on a black and white TV screen, for they have not been endowed with rods and cones in their retinas to change electromagnetic wavelengths into colorful magnificence.

How do objects acquire their colors? Why does the leaf appear to be green and the apple red? It turns out that this is because the atoms and molecules of most materials absorb some, and not all the visible wavelengths that fall on them. Which ones they absorb depends on their structure and constitution. Atoms and molecules have their characteristic tastes for waves, as it were. If a material sucks in every component of light and reflects back nothing it appears dark, as is the case with charcoal, for example.

So this is how the colors of bodies arise: by virtue of the different absorbing characteristics of their component atoms and molecules. Contrary to appearances, colored bodies do not emit the corresponding colors. When white light shines on them, they simply return to the world that color unacceptable to them. An orange pigment is one whose atoms absorb all but orange light. Ultimately, aside from rods and cones retinal neurons the silent atoms are what cause the colors of bodies. It is all a question of vibrations, preferred and appropriate, characteristic or unwanted.

Speed of light

Swifter than arrow from the Tartar's bow.

- SHAKESPEARE (*Midsummer-Night's Dream*)

There was a time when the flight of an arrow was regarded as fast. The swift darting of hares and leopards were considered fast also. Where else could pre-technological human beings observe motions so swift?

Today speeding cars, bullet trains and supersonic jets are commonplace. Speeds of the order of a few hundred miles an hour do not impress us anymore. But such speeds would seem incredible to someone from the Middle Ages.

We should be more surprised to know about speeds in the physical world. The speeds of the sun and stars, air molecules and electrons in atoms are truly impressive. But the speed of light is unimaginable. Light covers three hundred million meters (186,000 miles) a second: It is hard for human minds to conceptualize speed of this magnitude. Yet this is the most common speed in the universe: electromagnetic waves pervade every nook and corner of the physical universe for more than anything else.

More impressive is the mind that measures it. By ingenious means human beings have measured precisely how fast light travels in empty space and in other media. From the 17th century when the first determinations were made by the Danish astronomer Olaus Roemer to our own times numerous techniques have been developed to unravel this root of perceived reality.

The speed of light is the fastest speed in the universe. No physical body, massive as stars or minute as electrons, no physical body can reach a speed equal to that of light. We may picture particles zooming with speeds very nearly equal to that of light, but never equal to it. This is one more of the "You can't do it" principles of 20th century physics.

More remarkable and at first blush unbelievable is another aspect of the velocity of light: It does not depend on the motion of the observer relative to the source of light. To grasp the significance of this bewildering fact, imagine you are riding your bike at 15 mph towards a car. If the car is approaching you at 60 mph, it will appear to be coming at 75 mph since every hour this is the amount by which the distance between the car and you is diminishing. Likewise, for an observer moving forward in the same direction as the car, the car's relative speed will be 35 mph. Only if you are stationary will the car seem to be coming at 60 mph.

But this commonsense calculation does not work with light. Replace the car by a light wave, and the cyclists by very fast moving rockets, and everyone will find light to be traveling with the same speed (*relative to oneself*)! There will be not an iota less in the measured speed simply because you are whizzing towards or away from the source. The speed of light is a *universal constant*, as one says in physics, completely independent of the state of motion of the observer.

Faster than light

There was a young lady named Bright.
Whose speed was far faster than light;
*   She set out one day*
*   In a relative way,*
And returned home the previous night.

- ARTHUR BUTLER

That nothing can move faster than light is part of Einstein's famous special theory of relativity. One may ask, So what if something goes faster than light? Well, if that happened, then something as absurd as the above limerick would follow.

In fact, you can and do see the past when we see a star light-years away. The light reaching our eyes started that many years ago. One way of theoretically looking into the past would be following: We may picture the image of a scene now transpiring before our eyes to be traveling through space. Light from yesterday's event has by now traveled trillions and quadrillions of miles. Light from events that happened a thousand years ago have moved a lot farther.

Suppose you take off on a rocket that moves much faster than light. Then you will be able to overtake the light waves from earth and arrive at a planet before light from a past event reaches it. From that vantage position you can see whatever happened in the past when the corresponding light comes there. This imaginary experiment indicates how faster than light motion is equivalent to witnessing past events.

Theoreticians explore even strange phenomena, if they do not violate known principles. This they have done in the context of faster-than-light situations too. They have been led to a rather intriguing situation here: In principle it is not impossible for something to move faster than light if that something does not slow down to a speed equal to that of light. Let us clarify this: In order to speed up ordinary material bodies we need to give them energy. It requires infinite energy to speed up a body to that of light. Likewise, if entities exist which are already moving with speeds greater than light, such entities will require energy to lessen their speed, and the amount of energy needed for this would be infinite if we wish to slow them down to the speed of light.

This is the conclusion one draws form an analysis of the mathematics of the special theory of relativity. When the idea was conceived, physicists gave such (imaginary) entities a name: *tachyons*. Tachyons are particles that (in theory) always travel faster than light and can never be slowed down to the speed of light.

V. V. RAMAN

Scattering

The soft blue sky did never melt
Into his heart; he never felt
The witchery of the soft blue sky!

- WILLIAM WORDSWORTH

Ever since human beings turned their gaze skyward, they have recognized the vast sky above and admired its azure tint. The blue becomes more pronounced when white cottony clouds sail gently through the air. Soon after sunset, when the darkness of night takes over, all the blueness disappears. Even if the full moon shines bright in heaven the sky at night is never blue.

Though poets and painters had taken note of a blue sky, it was only in the 19th century that we came to understand why it is so. The phenomenon is related to a property called *scattering*. When light waves fall on a smooth surface they are reflected. However, when they encounter very small particles, they bounce back every which way: scattered in all directions. Not all wavelengths of light are scattered to the same degree. The scattering depends on the structure and size of the scattering center. Consider a beam of white light hitting a molecule of oxygen or nitrogen. The blue component of this light is scattered while the red and orange ones go undeflected. This simple fact has a dramatic effect on the nature of the perceived world. When sunlight enters the earth's atmosphere, its bluish components are scattered by the oxygen and nitrogen molecules. That scattered blue reaches our eyes when we do not directly investigate the sun. John Tyndall reasoned this out in the 1860s, after studying the matter experimentally. Light-scattering accounts the blue of the oceans too.

If our atmosphere were made up of some other gases which had the property of scattering the green component primarily, we would be enjoying a green sky. Who can tell what our poets would be singing then!

The study of scattering has revealed the substratum of matter. In 1890 Hendrik Lorenz used the scattering of sunlight in the atmosphere to estimate the number of molecules in a volume of air. Scattering can occur in liquids too. In the late 1920s C. V. Raman showed that by a careful study of light-scattering in a liquid one can determine the structure of its molecules.

Polarization: An aspect not directly perceived.

Understanding and manipulating the polarization of light is crucial for many optical applications.

- EDMUND OPTICS

Waves are disturbances that travel from one point or region of space of another. For every wave, there is a direction of propagation, and a line of oscillation of the traveling disturbance. As we saw earlier in longitudinal waves the line of oscillation is the same as the direction of propagation. In transverse eaves the oscillations are perpendicular to the direction of propagation.

In transverse waves, vibrations can occur on any plane perpendicular to the line of wave propagating. If there is only one line on this plane along which the vibration can take place, the transverse wave is said to be *polarized*. For example, vibrations in a taut string tied at both ends to two walls will be transverse vibrations: they vibrate along a line perpendicular to the string. If, the string passes through a long slit in a cardboard, its line of vibration would be restricted along the slit: we would have a polarized wave on the string.

Being a transverse wave light can also be polarized, a property that was discovered in the 19th century. The normal human eye cannot distinguish polarized from unpolarized light, but the eyes of some birds and insects can to do this.

Light is invisible

There is in God - some say -
A deep, but dazzling darkness... - **HENRY VAUGHAN**

Light itself is invisible: we can never see a ray of light

passing somewhere in space. The effulgent beam of light spouting out from a luminous source that a movie company displays as its logo cannot be seen if that light was splashed into empty space; nor the beams of science fiction movies. Only when it strikes our retina do we become aware of light. There is ample sunlight in the night sky. But it is only when some of it bounces back from the moon or a planet do we see it, and in become aware of the moon and the planets.

Most of the things we see is because of scattered light. The light we have in the room when glass windows are open and the light we experience in shade under a tree are there because light is scattered by the air molecules. Take away the air, and sunbeams will illumine only the patch on which they fall. We can only see reflections from objects. One seldom recognizes the importance of scattering.

It is like being on a volley-ball court. If you are not there, you will never receive the ball. But if you are there, the ball will come to you now and again, as and when the other players "scatter" the ball in your direction.

Thus we come to this unexpected realization: bodies are visible to us, not simply because of the light that falls on them, but equally because of the air around! Take away the light, and nothing can be seen. Take away the air, and not everything can be seen. Things will not be as visible in a room on the moon even in broad daylight because there is no air there.

This is one example of many that reveal how intertwining factors – often superficially unconnected - give rise to the world of perceived reality. Ours is a very much interconnected world. Only thorough and careful investigations slowly uncover those interconnections which are at the root of perceived reality.

Spectroscopy

Twinkle, twinkle, little star,
How I wonder what you are!   - JANE TAYLOR

Jane Taylor's legitimate wonderment, still echoed by little children to a happy and rhythmic tune, was answered a few decades after she passed way, thanks to the eagerness of physicists to find out all about light.

The first stone on the road to this exploration was laid by the lens grinder Joseph von Fraunhofer who constructed some excellent prisms. He attached them to a telescope and studied sunlight through them. He discovered in 1814 that the analyzed light from the sun displayed bright lines and dark. In the decades that followed, other investigators explored the phenomenon, and it was established that light from any source, when analyzed through a suitable optical device, gave a pattern of colors, of discrete lines or broader bands or continuous patches. The pattern of light seen through a prism is called a spectrum. The spectrum of light from a source is characteristic of the chemical composition of the source. It is a sort of fingerprint of the chemical elements present in the source of light. Soon it became clear that the spectrum can tell us also about the temperature of the source, and even about the source's motion relative to us.

These discoveries opened a new arena for physicists. Think of this: We get light from the sun and stars, and that light can tell us about the constitution of matter of which they are made! Just analyze the light from a distant source, and like a letter from a friend, you can know a good deal about the state and substance of the source. Soon scientists catalogued the spectra corresponding to various elements.

There is a related story that is interesting. Jules Janssen traveled to many places for the cause of science. In 1868 he was in India for a total solar eclipse to study the solar prominences which are particularly visible during eclipses. He was puzzled by a strange line in its spectrum. This was like finding the fingerprint of an individual, not in the police records. He sent his finding to Joseph Lockyer, an expert on solar spectra. After careful study of the line, Lockyer concluded that this must be a new element, hitherto unknown to terrestrial scientists. He called it *helium* in honor of the sun (In Greek *helios* means sun). If detective stories are fascinating, this one can beat any, Consider the tortuous route: a lens-grinder recognizes the spectra of elements in the 1820s, an astronomer discovers a new element in the sun in the 1860s! The existence of helium, the gas we use to fill birthday balloons, was first noticed in the sun during a solar eclipse observed in India. Light unraveled to human knowledge the existence of an element that is out there, 93 million years away! If this doesn't excite a mind, then of course we must read about the murders and rapes reported daily in our newspapers for thrills.

So we see how physicists have a way of finding out what the sun is made up of, and Polaris and Betelgeuse and whatever else in high heavens. It gradually became clear that stars and planets, high and mighty as they seem, are made up of the same sort of stuff as makes up this our modest planet here below. Aristotle was not quite right when he preached that celestial bodies consisted of incorruptible matter while earthly ones degenerate and decay. No, all bodies are created equal, though every piece of matter does not possess equal amounts of every kind of matter. Analysis and reductionism can give lots and lots of information.

Creation of light

...For mostly they go up and down
Or else goes round and round.

- PATRICK R. CHALMERS

We light a candle or a log, we flip a switch in the room or press the bottom on the flashlight, and light appears. And, of course, we have the sun which, at every rising, floods our surroundings with light. But, like the city kid who thought that the source of milk was the carton or the bottle, we would be mistaken to think that the source of light is the candle or the log, the light bulb of the sun.

Light is created by complex processes at the heart of matter. In the core of stars there is perennial transformation of matter into energy as per Einstein formula $E = mc^2$ mentioned before. Matter is transformed into insubstantial electromagnetic waves of different wavelengths. So, like aroma from brewing coffee, light emerges from the depths of stars as nuclear brewing go on. As will be seen later, processes producing light can also occur at lower temperatures.

It is also created in the heart of matter when electrons

in atoms jump from orbit to orbit. We will see more on this in another context.

Microwaves & application of scientific knowledge
I feel and seek the light I cannot see.

- SAMUEL COLERIDGE

Our eyes are sensitive to (i.e. we can see) only a very window of the electromagnetic spectrum. But this does not mean that others are insignificant or that we cannot know anything about them. One of the goals of science is to unravel every aspect of physical reality that can be perceived, either directly or indirectly.

Beyond red light, there are electromagnetic waves of longer wavelengths: the infrared (IR) or heat radiations. Beyond this are the microwaves. They have come to play an important role in technological civilization, but in the first third of our century they were not even named.

In the 1930s microwaves were used to determine the location and motions of distant objects using of radar. Like all inventions it has evolved to considerable complexity, used not only in planes taking pilots through thick and opaque clouds, but also helping ground controllers spot and guide them near airports. Microwaves are the carriers of TV signals and serve in long distance telephony as well. They penetrate through the reflecting layers in the upper atmosphere and are useful in communicating with astronauts in space. They have come to serve computers and medicine, and they also help us foods quickly.

In just a few decades after the discovery of microwaves, one started using them in many contexts. This is one major difference between the sciences of earlier centuries and those of our own times: the haste and lust to use every bit of scientific knowledge. Science is no longer just *natural philosophy*: love of knowledge of Nature. It is not even *science*: pure knowledge. It has become a tool for application and power, an instrument to exploit and control Nature. In this obsession to make life easier and comfortable, more enjoyable and materially fulfilling one often loses sight of the grander vision of science which is to understand and contemplate the complexities in the world, to marvel at the wonders of the cosmos. More seriously, rampant indiscriminate application of science has resulted in alarming dangers for the human condition

Radio astronomy
... he, repining not at lack of sight,
Might see as never man saw. **- RICHARD R. BOWKER**

Something very interesting happened in 1932. An

engineer by the name of Karl G. Jansky, while trying to solve the static problem (the hissing noise) associated with radio reception, made a significant discovery: that the earth is being showered from outer space by electromagnetic waves of wavelengths longer than the infra-red. We look at the sky and see the sun and stars: Light reaches us from the heavens. But day in and day out we are also inundated with invisible microwaves. If light can tell us so much about the sun and the stars, perhaps microwaves can tell us quite a bit too. Jansky laid the foundation for what was to become *radio astronomy*: the exploration of the universe, not with the aid of visible light, but by studying the microwaves that are continuously pouring in from every nook of the universe. On the mindless plane of physical reality, waves are just carriers of energy. In the world of perceived reality, they carry information too.

Radio-telescopes are very large dish antennas hooked to complex electronic circuits and computers which record and interpret the microwaves they detect. Many radio-telescopes have been constructed since the 1940s, and they are continually keeping a watchful eye round the clock on every sector of the skies, swallowing every surge of microwave that splashes on them. We have radio telescopes in Woomera in Australia and

ROOTS OF PERCEIVED REALITY

Arecibo in Puerto Rico, in Owens Valley in California and Gothenburg in Sweden, in Johannesburg in South Africa and Semiiz in Russia, in Fairbanks in Alaska and Jordell Bank in England, to name a few.

The radio-astronomical route has its advantage and disadvantage too. The advantage is that, unlike optical (light-based) astronomy, astronomers do not have to wait until it gets dark, nor for cloudless skies, to focus their instruments on stars and planets. Microwaves pouring in from the universe can be detected day and night, and they pass right through the clouds as light does through chunks of clear glass. The disadvantage is that we need much larger dishes (corresponding to lens sizes) to get clear pictures of whatever we are looking at. In other words, two sources will be blurred together when observed with even quite large (diameter a hundred meters) radio telescopes. One ingenious way by which radio astronomers have considerably increased resolution is by linking up radio telescopes in different parts of the globe and using the pairs as one single mammoth instrument. This technique is known as *Very Large Baseline Interferometry* (VLBI). Thus, telescopes in Bonn (Germany) and in Goldstone (California) team up and pinpoint on the same source and manage to get a much sharper reading.

Radio telescopes have enlarged our vision of perceived reality in many ways. They have put into evidence many sources of microwaves. Microwaves arise from supernova eruptions, electrons going amuck in interstellar magnetic field, and transitions in the atoms of hydrogen spread all over space. Though them we have detected carbon containing molecules like cyanogen and formaldehyde. This may suggest possibilities of organic molecules elsewhere in the universe.

Radio astronomy has revealed that elliptical galaxies emit considerably more radio waves than do most others. It has made us aware of thousands of incredibly powerful extra-galactic radio sources: some of the most fantastic objects in (from our perspective) the outskirts of our universe: mammoth star-like agglomerations spewing out incredible amounts of energy as they rush away at delirious speeds which are respectable fractions of the speed of light. These awesome things have been named *quasi-stellar objects* or *quasars* by radio astronomers. Their ultimate nature is still only faintly understood. Radio astronomy is like an extra window into the universe through which we have come to see many more wonders of perceived reality.

Many astronomers are convinced that somewhere out there among the billions of globules there must be other mind-endowed entities. Some of them may be more evolved than us in thinking and feeling, probing and inventing. If they are intelligent, they must have their radio astronomers too: sending knock-knock signals and expecting answers. So we need to be on the lookout for mail from extra-terrestrials, not in hard-copy formats, but in cryptic coding of microwaves. So eager radio-astronomers have been spending countless hours and (lots of dollars), not as peeping Toms, but as seekers of cosmic pen-pals. Whether we succeed in this lofty quest is not as important as the fact that human ingenuity has come up with tangible ways of confirming whether there are interstellar brain-based efforts. Like prayer, irrespective of whether it reaches a target, the effort itself enhances the human spirit.

Background radiation

Where did you come from, baby dear?
Out of the everywhere into here.

- GEORGE MACDONALD

According to a Chinese legend a celestial architect by the name of P'an Ku, born of the Cosmic Egg, worked hard for eighteen thousand years to build this grand universe. The ripples of this momentous event are still there: P'an Ku's breath and sighs we see to this day as winds and rising clouds; the roaring majesty of his voice resounds as thunder. His flesh congealed as earth, on which we can still feel his lush hair as green grass and tall trees. Metals and minerals underground are vestiges of his bones, while the abundant sweat of his lasting labors still drip down as rain. HJ too had lice infecting his bodies, and they may still be looked upon as swarms of humans populating the earth.

This picturesque fable explaining the origin of the universe underscores the idea that things we perceive today are long-range effects of primordial events of immense complexity.

In 1966, the radio astronomers Arnold Penzias and Bob Wilson discovered precisely that: a microwave radiation of wavelength 7.35 cm that is all pervasive and isotropic. They tried hard to see if this utterly uniform background radiation was perhaps due to local effects, like glitches in telescopes or noises in the atmosphere. But after every precautions and after every other reasonable possibility had been eliminated, they came up with conclusion: They had observed   a remnant of the world-generating Big Bang. Among other things, the enterprise of radio astronomy has put into evidence what may well be described as the first shriek of cosmic birth!

Gamma rays

Gamma rays are the electromagnetic radiations accompanying nuclear transitions.

- ROBLEY D. EVANS

At the lower extreme are the *gamma rays* which are so short in wavelength that atomic dimensions are large compared to them. This implies even more mind-boggling frequencies than light: of the order of 10^{22} Hz. This is a number beyond the grasp of normal people, for it represents ten trillion times a billion vibrations a second. If a magical machine were to print out numbers at the rate of one a second, and if that machine had started functioning when the Big Bang burst forth, then by now the machine would have reached only up to 10^{17}. The number corresponding to the gamma ray frequency is a hundred thousand times this!

Just as light emerges from electronic transitions in atoms, gamma rays arise when atomic nuclei squiggle, an electromagnetic expression for nuclear agitations. They are generated whenever there are nuclear explosions which civilized governments perpetrate for their supposed protection from potential enemies. These rays are like awfully penetrating bullets, going right through the thicknesses of ordinary matter. Should they encounter a living cell, they will simply shatter it. That is why in nuclear reactors thick concrete walls are built to absorb them as they emerge from the core. This destructive capability is utilized in annihilating cancerous cells in the body: That is what is involved in some radiation therapy. Gamma rays are also used in sterilization. Potatoes exposed to gamma rays do not sprout while waiting for customers at the supermarket.

In the 1970s astronomers detected that some of satellites launched for scientific purposes recorded gamma rays which seemed to be reaching us from heaven knows where. In fact, a satellite called CGRO (Compton Gamma Ray Observatory) was launched specially for the purpose of studying these in greater detail. In 1994 astronomers concluded from its data that way out there in very distant galaxies gamma rays are being produced in abundance by some strange processes. There is no limit to the activities in this fantastic cosmos, or to the bits of data we keep continually gathering about them.

Laser: light that doesn't spread out
The atoms become like a moth, seeking out the region of higher laser intensity   -STEVEN CHU

Ages ago human beings discovered fire, perhaps the very first light created by human ingenuity. Since then we have produced all sorts of light, both chemical and electrical, but invariably such light is dissipated: that is to say, like the flash from a torch-light it spreads all around, becoming weaker and weaker as it moves farther and farther away. We call this incoherent light. It consists of various wavelengths which move along different directions but are not quite in step. This is natural if we recall that light emerges every time an electron jumps to a higher orbit and falls back to a lower. Since the atoms are distributed at random and the "lifting up energy" also reaches them in a random manner. The resulting light is somewhat like the random exit of a crowd from a theater, everyone moving every which way.

That was the only kind of light known to us until at one time. Then in 1960 Theodore H. Maiman constructed a device, using a crystal of synthetic ruby, by which he created perfectly coherent light: i.e. light made up of identical waves in perfect step. Since the device caused *light amplification by simulated emission of radiation*, it received the acronym of laser. It was as if Maiman had arranged to have the crowd from the stadium walk in perfect step along a single path.

It started out as a toy, a contrivance to produce a narrow pencil of bright red light moving along a straight line like a long flashy arrow. One could use it in a lecture to point to a diagram or an important statement on the screen. But in less than two decades the invention found the most unexpected applications. Today they are used in compact discs and in checkout counters in stores; to clean paintings and in treating detached retinas; in communication systems and in detecting continental drifts. They have come into computers and are used in precise measurements: Thanks to lasers know the distance of the moon with an error of just one foot.

Consider the gigantic gush of water, noisily pouring down at Niagara Falls. Imagine for a moment that all this water can somehow be narrowed down to a very thin tube and made to hit the ground below. The spot would be struck with a tremendous force. So we can well imagine what a concentrated beam of powerful light can do. Lasers have been constructed with powers of barely a tenth of a watt. But there is one called Nova whose power is a hundred trillion watts. It is possible to shatter "a fuel tank removed from an intercontinental missile."

All the potential of lasers has yet to be realized. It is remarkable that human ingenuity has created a kind of light that, as far as we are aware, never existed before. We are both knowers and doers in this world.

When Thomas Gray, in *The Progress of Poesy* spoke of

blasted with excess of light, he did not imagine that it could someday describe the laser.

Photons

I give you the end of a golden string;
Only wind it into a ball.   **- WILLIAM BLAKE**

Thus far we have spoken about light as a wave that wings its way from point to point through electromagnetic vibrations. But light also behaves as if it is a volley of infinitesimally small specks of pure vibrations, carrying tiny bits of energy. Thus it is like miniature balls flying, while spinning furiously on an axis. Light (any electromagnetic wave) behaves as if it is made up of innumerable little bundles of energy moving with its customary speed of three hundred million meters per second in empty space. If we imagine a wave of light to be a golden string, then through one her marvelous tricks Nature winds it into a tiny ball. Expressed differently, there is also a particle aspect to light.

The particle aspects of electromagnetic waves are called *quanta* or *photons*. The energy carried by them is proportional to the frequency of the associated wave.

Particles and energy are different. One is localized, and the other spread all over. Yet, the fact is that light behaves as particle or as wave depending on the circumstance. A coin has both head and tail; throw it and only one side will show up when it falls to ground. Light is both wave and particle; do an experiment, and only one aspect will appear in the experiment.

People behave in a certain way in public, and very differently in the private. So too, at the normal scale of our experience light generally appears as a wave. At the microcosmic level of atoms and molecules, its photon aspect becomes predominant. This is one of the intriguing features of both matter and energy.

Effects of light

To have seen what I have seen, see what I see!

- SHAKESPEARE (*Hamlet*)

Light brings the far flung reaches of the universe together. Without it we will be earthlings condemned to perpetual isolation in a cold corner, and each one of us would have evolved in a darkness that would be as stifling as any self-centered existence can be.

Light is a major instrument in our interactions with the world. It informs us of the presence of persons and things beyond ourselves. It reveals their shapes and forms. It unveils their beauty. It is light that speaks to us of distant stars: of the nature and substance of unapproachable celestial entities. It tells us if a galaxy is receding and at what rate. Light is a silent messenger. It makes no noise yet carries enormous amounts of information.

It is not only at the intellectual level that light serves us. The chemical constitution of the star Antares may be interesting, and the distance of the Andromeda nebula may be impressive. But there is more to life than knowledge and wonder. Enjoyment is no less important. Here too light plays a role. For there is more to light than brightness. Light is not simply vibrations of varying intensities, but of satisfying shades and colors as well. Color adds splendor to the world.

Light has all the properties characteristic of waves. It is reflected and refracted, it is deflected and diffracted. It is indifferent to things not on its path and affects only to those that are on its way. The properties of light add to the charms of the visible world. The changing colors of the diamond beetle arise, for example, not from pigmentation, but because of diffraction. The glory of the rainbow and the colors of the icicle result from refraction. The blue of the sky is due to light scattering. Without light diamond would be as dark as a piece of charcoal, while rubies, sapphires and emeralds would all be dark as the depths of hell.

The effect of light on the world around us is of

incredible variety. The magnificent aurora and multihued butterflies, the poetry of flowers and the canvass of artists, all depend on light. No wonder, language itself has been enriched by light: a source of light may glimmer or glow, it may dazzle or shine, twinkle or radiate.

Light is a life-sustaining principle: It transports sun's energy to here below, and by cleverly collaborating with the green plants, it feeds that energy to living organisms.

As J. J. Thomson reminded us, "The study of light has resulted in achievements of insight, imagination and ingenuity unsurpassed in any field of mental activity..."